The Theory and Design of Cycloconverters

MONOGRAPHS IN MODERN ELECTRICAL TECHNOLOGY

ALEXANDER KUSKO, SERIES EDITOR

1. Solid-State DC Motor Drives by Alexander Kusko
2. Electric Discharge Lamps by John F. Waymouth
3. High-Voltage Measurement Techniques by Adolf J. Schwab
4. The Theory and Design of Cycloconverters by William McMurray

The Theory and Design of Cycloconverters

WILLIAM McMURRAY

The MIT Press

Cambridge, Massachusetts, and London, England

Library of Congress Cataloging in Publication Data

McMurray, William, 1929-
The theory and design of cycloconverters.
(Monographs in modern electrical technology, 4)
Bibliography: p.
1. Electric current converters. 2. Thyristors.
I. Title.
TK7872.C8M3 621.313'7 70–178121
ISBN 0–262–13081–5

Contents

Preface

The purpose of this book is to provide a comprehensive in-depth treatment of cycloconverters of the naturally commutated type so that accurate performance and design calculations can be made. A cycloconverter is essentially a pair of phase-controlled rectifiers connected in inverse-parallel, the output of which is modulated to provide ac power at a frequency generally lower than the source frequency. The modulation makes the analysis of these circuits rather difficult.

Applications include adjustable-speed ac motor drives including position servos and VSCF (variable speed, constant frequency) power supplies, as used in aircraft. There is interest in the development of electric propulsion for off-highway vehicles, using the chain: gas turbine, high-speed alternator, cycloconverter, induction motor, gear train and wheels. The first three links of this chain can provide a lightweight general-purpose power supply, which is of interest to the military. These modern applications use solid-state thyristors (silicon controlled rectifiers) as the power switching devices. Some initial work was done in Germany during the 1930s using mercury-arc tubes to convert 50 Hz to $16\frac{2}{3}$ Hz for electric railroads.

Although much development work has been done, the practical application of cycloconverters has been restricted by the cost of the equipment. At the time of writing, there was no published information of any depth in books or technical papers on cycloconverters. The available information was scanty, scattered, and omitted detailed derivations. The lack of a theory sufficiently detailed to allow accurate performance and cost projections has contributed to the slowness of development. However, I believe that cycloconverters are now on the verge of an expansion in application, and a book on this subject is timely.

Since the writing of this work, a book by B.R. Pelly has been published (Reference 9) that gives an excellent coverage of cycloconverter theory. However, the emphasis of Pelly's work differs considerably from the treatment presented here, and there is surprisingly little duplication, considering the similar subject matter. Some aspects of cycloconverter operation that are lightly covered by Pelly receive more depth of study here, and vice versa.

This book is a result of theoretical studies of cycloconverter operation for several types of application. The analysis is based on the well-known "continuous current" theory of the phase-controlled rectifier with relatively low-frequency amplitude-modulation and the derivation of a Thévenin equivalent circuit. It is assumed that the reader is familiar with the fundamental theory and practice of phase control and also with the basic principles of cycloconverter operation.

This work is intended as a reference for engineers working in the field and graduate students specializing in electric power conversion and control. It could be used as collateral reading for graduate, postgraduate, or on-the-job courses in electric power conversion, adjustable speed drives, aerospace power systems, or electric vehicles. However, the scope is too narrow for a text, unless cycloconverters were to achieve sufficient importance to justify a separate course. Some engineers interested only in phase-controlled rectifiers with dc output will find certain portions of the book to have reference value.

Acknowledgment is expressed to my colleagues in General Electric Corporate Research and Development, to members of the staff of the Industrial Semiconductor Products Business Section of General Electric Semiconductor Products Department, and to the Mobility Equipment Research and Development Center, U.S. Army, Fort Belvoir, Virginia, for its support of portions of the work described in this text. The author also wishes to thank R. A. Hamilton and G. R. Lezan of the G. E. Custom Power Equipment Department and D. L. Plette of the Aerospace Electrical Products Business Section of the G. E. Aircraft Equipment Division for application information used in Chapter 11.

William McMurray

1 Introduction

Phase-controlled rectifiers are quite well known, and there are a number of books on the subject.[1–9] The modulation of such converters to produce ac ouput, while simple in principle, greatly complicates their quantitative performance. An attempt is made here to treat various effects of the modulation in theoretical detail and show how the resulting equations can be used for accurate performance and design calculations. The work is oriented toward general circuit design rather than any specific application.

This analysis of the cycloconverter is based upon the classical equations of a phase-controlled rectifier operating with continuous load current. These equations are quoted for reference, but no derivation is given. The general algebraic form of the equations, applicable to many different circuit configurations, is retained throughout the the work, so that most practical circuit arrangements are covered as particular cases. A complete glossary of the mathematical symbols employed and standard definitions of rectifier circuit terminology are included at the back of the book. The calculation of many cycloconverter performance factors requires numerical integration of corresponding rectifier equations, and the results are presented in tabular and graphical form. Some of the analyses would be impractical without the aid of a computer.

This book brings together, under one cover for ready reference, a comprehensive study of various aspects of cycloconverter behavior. The following included topics have not been analyzed in depth in previously published literature:

Detailed equivalent circuit, including frequency ratio correction factors.

Theoretically based criteria for the quality of the output waveform as a function of frequency ratio, pulse number, and modulation fraction.

Performance with nonsinusoidal modulation.

Computation of input power factor considering the effect of overlap during commutation.

Quantitative data on the harmonics of the supply frequency in the output voltage.

Calculation of the losses in snubber circuits connected across thyristors or between phases.

Effect of thyristor rating parameters on the design of cycloconverters.

This work extends the analysis to about the limit of practical usefulness for this approach. The assumptions and limitations of the treatment are pointed out. If more accurate performance prediction is required, then a direct simulation of the circuitry on an analog, digital, or hybrid computer is necessary. Such simulation is beyond the scope of the book.

The subject matter is also restricted to converters that are commutated naturally by the ac supply voltage. An excellent treatise on force-commutated cycloconverters has been written by Laszlo Gyugyi.[58] However, the additional equipment needed for force commutation makes this type of circuit more expensive, and it is not likely to find practical application in the immediate future.

1.1 CIRCUIT CONFIGURATIONS AND ASSUMPTIONS OF THE ANALYSIS

The equations have been generalized to apply to a class or family of converter circuits, namely simple p-phase single-way or double-way circuits, where $p \geqslant 2$. As used here, p is the number of supply phases in each commutating group, termed the "commutating number," while q is the "pulse number" of the complete circuit, which determines the harmonic content in the output. This family includes the common 3-phase star (wye or single-way) circuit in Figure 1.1 and the 3-phase bridge (double-way) in Figure 1.2. Series or cascade connections of such circuits are also included, but interaction through common source impedance is neglected. It should be quite easy to reduce other converter configurations (such as parallel or multiple connections) to an equivalent form.

It is assumed that the source is a stiff, sinusoidal, balanced polyphase system. The line inductances L_p may consist of transformer leakage inductance, supply system inductance, alternator inductance or inductance specifically added to control the behavior of the cycloconverter. If other converters (such as those producing other output phases) are operated from the same supply, it is assumed that they are provided with substantially separate sets of inductances L_p so that interaction between the converters can be disregarded. Saturating reactors SR may be connected in series with each thyristor, if desired, to limit the initial di/dt and to reduce the size of the

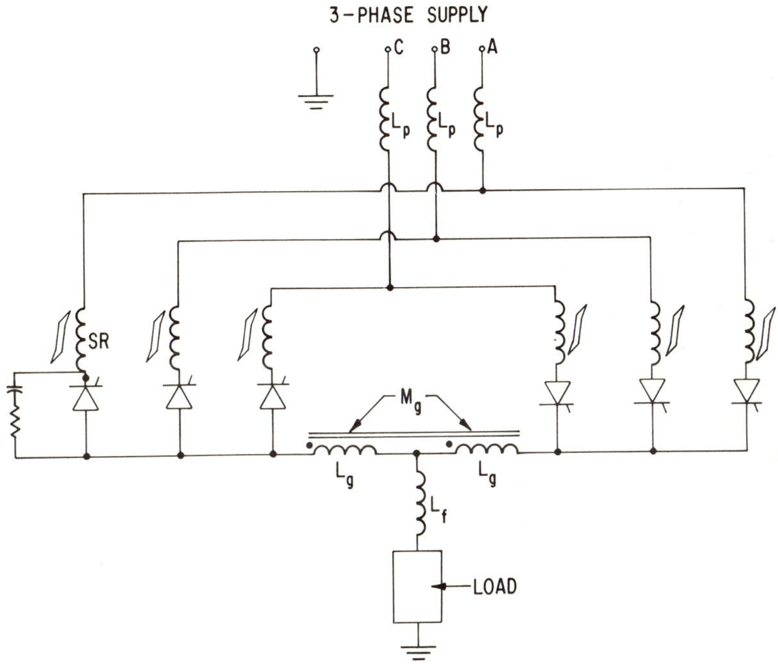

Figure 1.1. 3-Phase single-way cycloconverter.

RC snubber circuits (shown across one thyristor in Figure 1.1) that are required to limit dv/dt and supress voltage spikes.

In the "forward-reverse" type of cycloconverter, the group of thyristors that conduct positive load current can be separated from the group that conduct negative load current by a center-tapped intergroup reactor that will limit the current that may circulate between the two groups. Each half has a self-inductance L_g, and the two halves are coupled by a mutual inductance M_g, as indicated in Figures 1.1 and 1.2. Note that intergroup reactors do not behave as interphase transformers in multiple converter circuits, and the two functions cannot be performed by windings on a common core. If the intergroup reactor is omitted, the pairs of thyristors can be connected directly in inverse-parallel, sharing a common RC snubber and saturating reactor (if used). An output filter inductance L_f may be connected in series with the load, if necessary, as when the load has a capacitor input filter or is a stiff counter emf.

The "delta-rectifier" type of cycloconverter, consisting of three rectifier converters connected in delta, has no reverse polarity groups

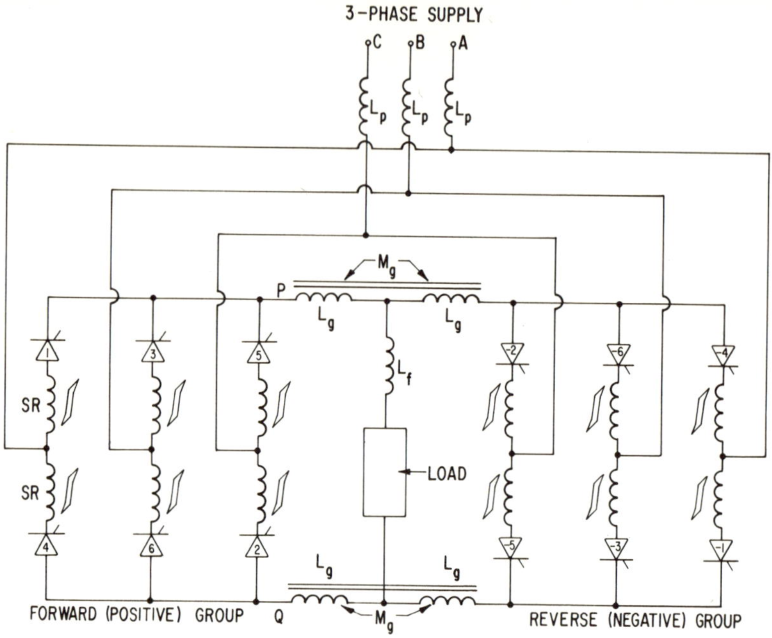

Figure 1.2. 3-Phase double-way cycloconverter.

as such, and therefore no intergroup reactors. This type of circuit is treated as a special case in Chapter 9.

The analysis assumes that the load current is sinusoidal, of a given magnitude and phase angle. This is believed to be the most significant limitation of the theory presented here, particularly when no filter reactors L_f are employed, as in most motor drives. Distorted current waveforms increase the losses and reactive power of the converter above the values predicted by the idealized theory.

1.2 THYRISTOR FIRING CONTROL

Implicit in much of the analysis are certain assumptions regarding the method by which the cycloconverter is controlled. The transfer function between the input control signal and the ideal output voltage is linearized by a phase control method that employs cosine "firing waves" unless a different control method is specified. The ideal output voltage is the pseudo-instantaneous output voltage neglecting commutation overlap, losses, the effect of nonlinear reactors SR, and frequency ratio distortion. Pseudo-instantaneous quantities are defined as values averaged over the interval between the firing of one thyristor and the next.

Examples of the actual instantaneous output voltage waveform of a 6-pulse cycloconverter are presented at the top of Figures 1.3 and 1.4, showing lagging and leading power factor load conditions, respectively, and the production of these waveforms by the control method using cosine firing waves is graphically illustrated in the lower parts of the figures. The output voltage is denoted by the heavy line superimposed upon the set of possible instantaneous output voltages for a 6-pulse converter. These are the line-to-neutral voltages of a 6-phase star circuit or the line-to-line voltages and their inverse for a 3-phase bridge, such as shown in Figure 1.2. The converter

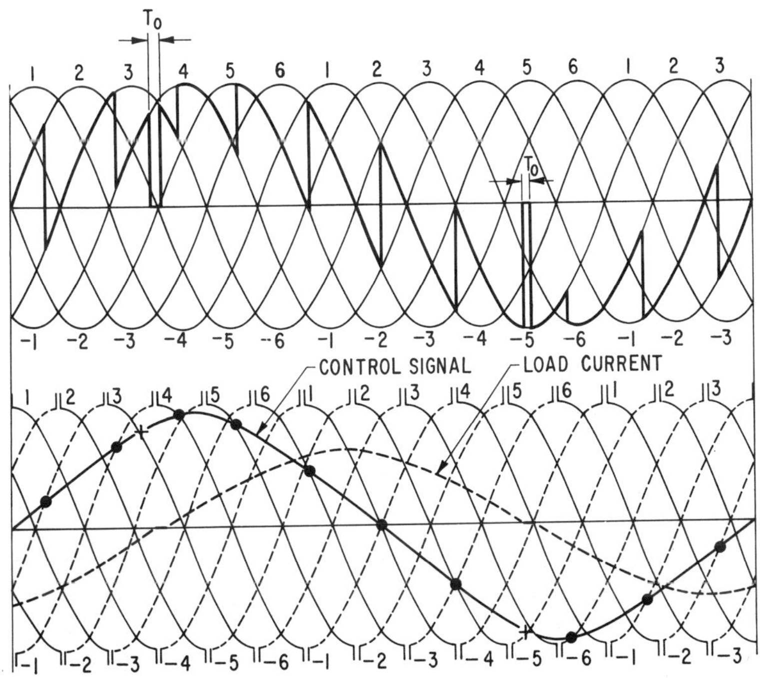

Figure 1.3. Construction of 6-pulse cycloconverter output voltage, lagging power factor load.
Top: Output voltage and set of possible instantaneous output voltages.
Bottom: Cosine firing waves, control signal, and assumed load current. Thyristor firing instants are the dotted intersections.

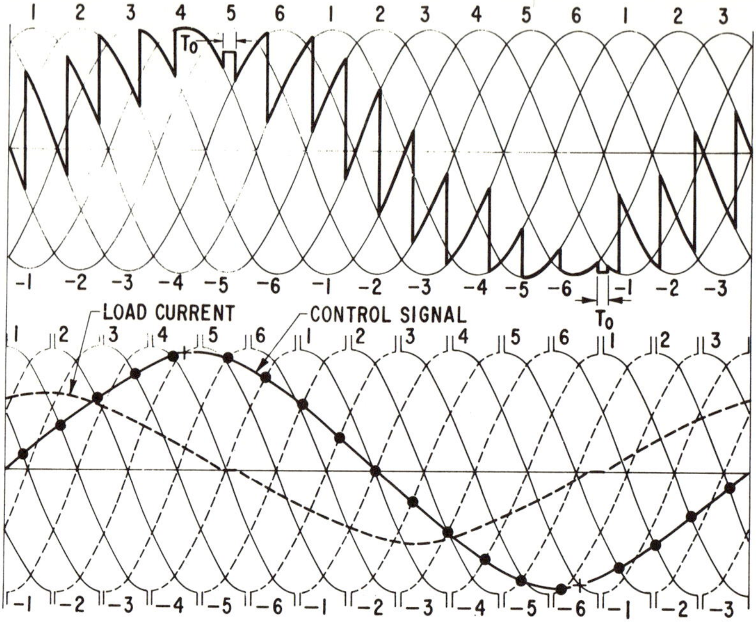

Figure 1.4. Construction of 6-pulse cycloconverter output voltage, leading power factor load.
Top: Output voltage and set of possible instantaneous output voltages.
Bottom: Cosine firing waves, control signal, and assumed load current. Thyristor firing instants are the dotted intersections.

output voltage during conduction of a particular thyristor having a firing sequence number indicated in Figure 1.2 follows the correspondingly numbered voltage wave in Fig. 1.3 or 1.4 (positive numbers for forward group output, negative numbers for reverse group output).

The corresponding cosine firing waves for both forward and reverse polarity groups are shown below, distinguished by solid and dashed lines, respectively. The firing instants of the thyristors are determined by the intersection of the control signal (the solid-line low-frequency sine wave) with the cosine waves. An output/input frequency ratio of 0.4 is shown. Note that the firing wave for each thyristor is restricted to a range of 180° or less, the maximum normal firing angle variation, and is bounded at either end by vertical lines that represent the firing angle advance and retard limits that are included in the control circuits. If the control signal magnitude exceeds the peak of the cosine wave, it intersects these vertical lines, and the firing angle is held against the limit. This results in "saturation" of the transfer function between the control signal and the output, or "clipping" of a periodic waveform. The segregation of positive and negative group waveforms in Chapter 5, Figures 5.1 and 5.2, may aid in clarifying the construction of Figures 1.3 and 1.4.

An assumed sinusoidal load current is indicated by the dashed-line low-frequency wave superimposed on the control waves. The implied method of eliminating circulating current is to employ "blanking" or suppression of firing signals to the negative group while the positive group is conducting current and vice versa. Thus, when the current is positive, a positive group thyristor is fired when the control signal intersects the cosine wave for that particular thyristor, and the output voltage switches accordingly. Similarly, negative group thyristors are fired only when the current is negative. The intersections that result in a thyristor firing and output waveform step are dotted in Figures 1.3 and 1.4.

In the usual cycloconverter control scheme using two separate control channels for the forward and reverse thyristors, the control circuitry of the inactive group maintains a continuing record of the proper firing angle for said group, which would produce the same pseudo-instantaneous output voltage as actually provided by the active group. This assumes the continuous-current mode of operation and no distortion of the natural supplementary relationship between the forward and reverse firing angles (that is, their sum is 180°). The inactive thyristors should be disabled by locking out their gate pulse amplifiers and not by biasing their control circuit to the retard limit. Then, when a current reversal is required and the outgoing group is locked out at current extinction, the incoming group can be released immediately after the turn-off delay T_0. Assuming that gate pulses commence as soon as the lockout is released, the thyristors fired are those that would already be conducting had the current always been in the new direction. This action generally provides a transition having minimum disturbance.

With passive inductive load, the output voltage during the deadband T_0 falls to zero (see Figure 1.3). The thyristor that fires at the end of this interval is the one that would have fired at the intersection indicated by a cross had it not been locked out at the time. When the load has a leading power factor, as in Figure 1.4, the output voltage during the deadband T_0 dwells at the load counter emf, which is needed to cause the assumed current extinction. The effect of this deadband required to allow recovery of the outgoing thyristor is neglected in the present analysis.

The deadband can be reduced or eliminated by applying firing signals to both groups of thyristors at all times. However, large intergroup reactors are usually required to reduce the resulting circulating current.[57] Sometimes, as a compromise, dc bias signals are added to reduce or eliminate circulating current by advancing the firing of the positive group thyristors and vice versa, but some deadband distortion is reintroduced.

Other defects related to the control technique can cause distortion of the output. If the cosine firing waves have a phase error with respect to the supply line voltages, "magnitude cosine" distortion results.[33] If they are obtained from a point on the supply lines where commutation notches appear, then undistorted signals must be carefully reconstructed by adding signals derived from current transformers with an appropriate burden.

The converter must enter the discontinuous current mode of operation if the load is sufficiently light; this will alter the circuit behavior, but this need not be considered in the design and rating calculations where the heaviest load is of prime concern. If the control circuits for the positive and negative groups are not closely matched, then direct current may flow in the load. All of these imperfections that may be present in the method of control are ignored in the present analysis. In practice, the problem can be mitigated and the linearity of the transfer function improved by appropriate feedback or other compensation techniques.[29]

1.3 POLYPHASE OUTPUT

Each phase of a polyphase load requires a separate converter that must be suitably isolated from the others. This may be achieved by using separate supply transformers for each converter, or isolated secondary windings on a common set of transformers, or isolated generator windings for that type of source. However, the power flow in a single-phase converter fluctuates widely at the second harmonic of the output frequency, so that the utilization of isolated supply windings is poor. With n_0 output phases, the total power flow from source to load is much smoother, the magnitude of the fluctuations being greatly reduced and occurring at the $2n_0$th harmonic of the output frequency. Thus, common input power lines for all output phases are preferred, unless the isolating reactors L_p can be incorporated in the separate transformer or generator windings. A design that segregates the leakage fluxes as much as possible is desired.

If the input lines are common, then isolation must be provided on the output side. For a motor load, separate windings for each phase can be provided quite easily. Ouput transformers are more undesirable than input transformers, because they operate at a lower frequency and are therefore larger. An advantageous arrangement is to use simple single-way converters that require no isolation at either the input or output side.

Provided that the proper input/output isolation is maintained, the output connection of a polyphase cycloconverter can be any one of the familiar arrangements that are used for transformers. For example, wye, delta, open delta, and zigzag connections are all feasible. Many such arrangements are shown by Pelly,[9] but they all employ simple single- or double-way converters as building blocks. The load is reflected into the individual simple converters in the same manner that the load on a polyphase transformer is reflected into its individual windings. Since the utilization factors of transformer windings are well known, the similar utilization factors of individual cycloconverters in polyphase connections will not be discussed here. Once the loading on an individual simple converter has been determined by these factors, the circuit may be analyzed and designed by the methods presented in this book.

The desired sinusoidal output is normally obtained by sinusoidal modulation of the input control signal. However, if 3-phase output converters are wye connected, then a certain amount of the third harmonic component can advantageously be added to the modulating signal. This component is eliminated from the output line-to-line voltages and from the load phase voltages provided that the load neutral is isolated from the supply neutral, between which that harmonic component appears. The amount of harmonic modulation can be selected to maximize the ratio between the fundamental and the peak output voltages of each converter. This improves the utilization and power factor of the equipment. In practice, the modulating signal can have a trapezoidal waveform, which is generally easier to produce.

2 General Equations of a Phase-Controlled Rectifier

Since the naturally commutated cycloconverter is, actually, a special type of phase-controlled rectifier/inverter, a thorough understanding of the behavior of such ac/dc converters is essential for the detailed study of cycloconverters. However, it is not the purpose of this book to repeat the fundamental analysis of ac/dc converters, a subject that is readily available in a number of textbooks[1–9] and technical papers. Instead, a brief review of converter waveforms and certain parameter definitions will be presented. Then, the general equations describing the performance of a phase-controlled rectifier or inverter will be quoted without derivation, both as a reminder to the reader and for reference during the analysis of cycloconverters in later sections.

Using the 3-phase bridge circuit as an example, Figure 2.1 shows typical waveforms produced by a phase-controlled rectifier (A) and a phase-controlled inverter (B). Such waveforms might be generated at different times during the operation of the cycloconverter in Figure 1.2 by the forward group of thyristors 1 to 6 at the left of that figure. The voltages at poles P and Q indicated on Figure 1.2 follow the waveforms shown in Figure 2.1 with respect to the neutral point of the ac supply system. In Figure 2.1, the reference line X to X is the time of natural commutation of pole P from line C to line A; that is, the time when continuous current to the load from pole P would begin to switch from line C to line A if thyristors 5 and 1 were ordinary diode rectifiers or were fired as early as possible. It is the time when the voltage of line A becomes more positive than line C.

The basic property of phase control is to delay the start of commutation by a phase angle α measured from the natural point. The angle designated α in Figure 2.1 is the firing delay angle for thyristor 1. For steady-state dc output, the firing delay angles of the other thyristors 2 to 6 have the same value α. In a cycloconverter, the firing delay angle is modulated at the ac output frequency.

Because of inductance, commutation cannot occur instantaneously but requires an interval of phase duration u. This is termed the commuting angle or angle of overlap. The angle designated u in Figure 2.1 is the interval in which conduction of thyristors 5 and 1

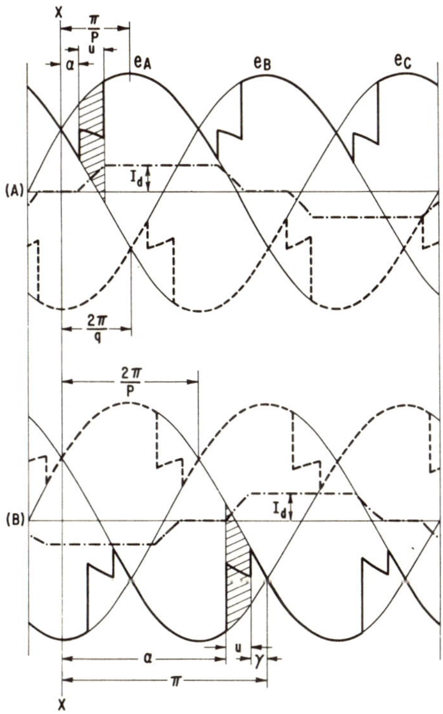

Figure 2.1. Typical waveforms in a 3-phase bridge ac/dc converter.
(A) Rectifying mode. (B) Inverting mode.
Solid lines: Voltage of dc pole P.
Dashed lines: Voltage of dc pole Q.
Chain lines: Current in phase A line.

overlap while current commutates from line C to line A. The voltage difference between lines A and C during this time (the shaded area in Figure 2.1) appears across the total commutating inductance, which is the inductance L_p in line A plus the inductance L_p in line C and the inductances of any reactors SR in series with the individual thyristors 1 and 5. The voltage is termed the commutating voltage e_c, which forces the change of current in the inductances.

During the commutation overlap interval, the output pole voltage follows the average of the incoming and outgoing lines. This causes the output voltage to be less positive (rectifying) or more negative (inverting) than it would otherwise be for the same delay angle α. For a successful commutation in the inverting mode, current transfer must be completed before the commutating voltage reverses, which would reverse the direction of current change. Since the voltage e_c is a sine wave starting from the point of natural commutation, it reverses after an angle π. Hence, the firing angle α must be sufficiently less than π to allow for the commutating angle u and a margin angle γ, which must be greater than the turn-off time of the thyristors. The margin angle γ for thyristor 1 is indicated on Figure 2.1(B).

The angle between successive commutations within any one commutating group of thyristors is $2\pi/p$, while the angle between successive commutations of any thyristors in the converter is $2\pi/q$. These angles are also indicated on Figure 2.1, where $p = 3$ and $q = 6$ for the 3-phase bridge circuit. In the equations that follow, the firing delay angle α and the dc load current I_d are considered to be the independent variables. The number of commutating groups connected in series or cascade is s. The 3-phase bridge circuit is effectively a cascade connection of two 3-phase commutating groups, for which $s = 2$. (See Glossary for further definitions.)

2.1 AVERAGE COMPONENTS OF THE OUTPUT VOLTAGE

dc Voltage without firing delay or commutation overlap,

$$E_{do} = E_m \, s \frac{p}{\pi} \sin \frac{\pi}{p}. \tag{2.1}$$

dc Voltage with firing delay α, but no overlap,

$$E_{d\alpha} = E_{do} \cos \alpha, \tag{2.2}$$

where $0 < \alpha < \pi$.

dc Voltage drop due to commutation overlap,

$$E_x = s\,pf(L_c + L_u)I_d \qquad \text{if} \quad I_d \leqslant I_s; \tag{2.3}$$

$$E_x = s\,pf(L_cI_d + L_uI_s) \qquad \text{if} \quad I_d > I_s, \tag{2.4}$$

where $L_c = L_p + L_s$.

dc Voltage drop due to equivalent resistance of R_c ohms/phase,

$$E_r = s\,R_cI_d.$$

dc Voltage drop due to thyristor conduction drop U_a volts, or arc drop in the case of gas tubes (this is the approximately constant component not included in R_c),

$$E_a = s\,U_a.$$

dc Voltage with firing delay, overlap, and losses,

$$E_d = E_{d\alpha} - E_x - E_r - E_a. \tag{2.5}$$

Except for very light loads, it may be assumed that $I_d > I_s$ so that Equation 2.5 may be written in detail as

$$E_d = E_{do} \cos \alpha - spf(L_u I_s + L_c I_d) - s R_c I_d - s U_a. \tag{2.6}$$

The terms of Equation 2.6 can be rearranged in the following form, for which the equivalent circuit shown in Figure 2.2 can be drawn:

$$\begin{aligned} E_d &= E_{d\alpha} - \text{(constant drop)} - \text{(drop proportional to current } I_d\text{)} \\ &= E_{d\alpha} - E_b - R_0 I_d. \end{aligned} \tag{2.7}$$

The second term is an equivalent back emf,

$$E_b = spfL_u I_s + s U_a. \tag{2.8}$$

The coefficient of the last term is defined as the equivalent output resistance R_0,

$$R_0 = spfL_c + s R_c. \tag{2.9}$$

Note that only the component $s R_c$ is a real watt resistance.

The angle of commutation overlap u is related to the drop E_x and the firing delay angle α by

$$\cos \alpha - \cos(\alpha + u) = 2 \frac{E_x}{E_{do}}, \tag{2.10}$$

from which u may be calculated, *

$$u = \text{Cos}^{-1} \left[\cos \alpha - 2 \frac{E_x}{E_{do}} \right] - \alpha. \tag{2.11}$$

If the drops E_r and E_a are neglected, the dc voltage E_d is also given by

$$E_d = E_{do} \frac{\cos \alpha + \cos(\alpha + u)}{2}. \tag{2.12}$$

The above equations and equivalent circuit are valid so long as each commutation is completed before the next begins,

* Throughout this work, the convention of capitalizing inverse trigonometric functions to denote the principal value will be used where only that value is valid in the equations. The principal value of the inverse cosine is restricted to the range 0 to π radians. For the inverse sine and tangent, the range is $-\pi/2$ to $+\pi/2$ radians.

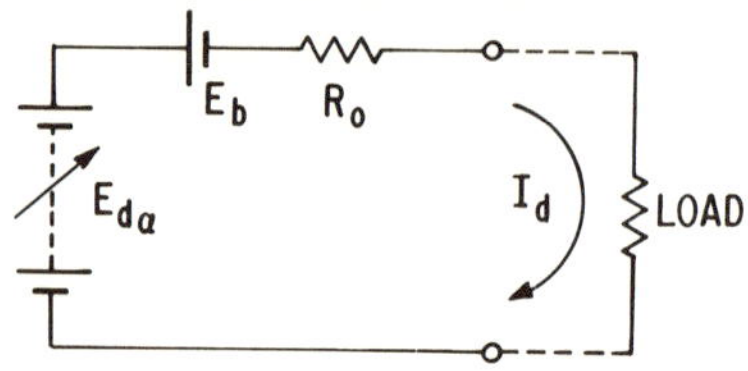

Figure 2.2. Equivalent circuit of a phase-controlled rectifier.

$$u \leqslant \frac{2\pi}{q}. \tag{2.13}$$

This range is termed the region of mode 1 or regular operation. Note that in double-way circuits with all the commutating inductance in individual legs and none in the common lines ($L_c = L_s$; $L_p = 0$), then p instead of q should be used in Restriction 2.13.

The equations are valid for both rectifier and inverter operation provided that a margin angle γ remains, during which reverse voltage is applied to the thyristors, to allow them to recover forward blocking ability.

$$\gamma \geqslant 2\pi f t_0,$$

where $\alpha + u = \pi - \gamma$ or

$$\cos(\alpha + u) = -\cos\gamma. \tag{2.14}$$

Thus, with load current I_d, the greatest negative voltage with which the circuit can operate as an inverter is

$$-E_d = E_{do}\cos(2\pi f t_0) - (R_0 I_d + E_b).$$

2.2 RIPPLE COMPONENTS OF THE OUTPUT VOLTAGE

In this section, it is convenient to express voltages as a function of the peak instantaneous output voltage E'_m, in terms of which Equation 2.1 becomes

$$E_{do} = E'_m \frac{q}{\pi} \sin\frac{\pi}{q}. \tag{2.15}$$

For simple single-way circuits, $E'_m = E_m$, of course. An angle β is defined by

$$\sin\beta = \frac{E_d}{E'_m}.$$

The time integral of the ripple voltage, which gives the ac flux linkage required by the dc filter choke L_f, is

$$\int e\,dt = \frac{E'_m}{\omega}\left[2\cos\beta - (\pi - 2\beta)\sin\beta\right] \tag{2.16}$$

if

$$\alpha + u \leqslant \beta - \frac{\pi}{2} + \frac{\pi}{q};$$

or

$$\int e\,dt = \frac{E'_m}{\omega}\left\{\cos\left(\alpha + u + \frac{\pi}{2} - \frac{\pi}{q}\right) + \cos\beta - \left[\frac{\pi}{q} + \frac{\pi}{2} - \beta - (\alpha + u)\right]\sin\beta\right\} \tag{2.17}$$

if

$$\alpha + u \geqslant \beta - \frac{\pi}{2} + \frac{\pi}{q}.$$

The maximum value of this integral is obtained when $u = 0$ and $\alpha = \pi/2$, for which $E_d = 0$ and $\beta = 0$,

$$\left[\int e\,dt\right]_{\max} = \frac{E'_m}{\omega}\left[1 - \cos\frac{\pi}{q}\right]. \tag{2.18}$$

The peak-to-peak current swing is

$$\Delta I = \frac{1}{L_f}\int e\,dt. \tag{2.19}$$

In the case of multiple rectifier circuits, the flux linkages required by each winding of the interphase transformer are also given approximately, but not exactly, by Equation 2.16 or 2.17 in which q is the pulse number of each of the component simple rectifiers.

The instantaneous output voltage, including harmonics, can be expressed as the Fourier series

$$e = E_d + \sum_{m=1}^{\infty}\left[a_m \sin mq\omega t + b_m \cos mq\omega t\right]. \tag{2.20}$$

When time t is measured from a point of natural commutation, the coefficients a_m and b_m are given by

$$a_m = \frac{E_{do}}{2}\left[\frac{\sin(mq+1)(\alpha+u) + \sin(mq+1)\alpha}{mq+1} - \frac{\sin(mq-1)(\alpha+u) + \sin(mq-1)\alpha}{mq-1}\right]; \tag{2.21}$$

$$b_m = \frac{E_{do}}{2}\left[\frac{\cos(mq+1)(\alpha+u)+\cos(mq+1)\alpha}{mq+1} - \frac{\cos(mq-1)(\alpha+u)+\cos(mq-1)\alpha}{mq-1}\right]. \quad (2.22)$$

If overlap is neglected ($u = 0$), these equations reduce to

$$a_m = E_{do}\left[\frac{\sin(mq+1)\alpha}{mq+1} - \frac{\sin(mq-1)\alpha}{mq-1}\right]; \quad (2.23)$$

$$b_m = E_{do}\left[\frac{\cos(mq+1)\alpha}{mq+1} - \frac{\cos(mq-1)\alpha}{mq-1}\right]. \quad (2.24)$$

The special case of a diode rectifier is obtained by substituting $\alpha = 0$ in Equations 2.21 and 2.22. If both α and u are zero, then $a_m = 0$ and

$$b_m = \frac{-2E_{do}}{(mq)^2 - 1}.$$

The resultant crest magnitude of (mq)th harmonic component is, for all cases

$$c_m = \sqrt{a_m^2 + b_m^2}.$$

2.3 THYRISTOR REQUIREMENTS

In the class of converter circuits considered here, the peak voltage applied to the thyristors is the maximum voltage between any two supply lines,

$$U_m = 2E_m \quad \text{if } p \text{ is even;}$$

$$U_m = 2E_m \cos\left(\frac{\pi}{2p}\right) \quad \text{if } p \text{ is odd.} \quad (2.25)$$

If the supply voltage is variable (including any regulation tolerance), then the highest value must be used for determining thyristor peak voltage rating. This neglects any commutation spikes or transients on the supply lines. Usually, a safety factor of 2 to 2.5 is used.

The rise and fall di/dt of thyristor current is proportional to the voltage difference between the incoming and outgoing phases, termed the commutating voltage e_c,

$$e_c = E_{cm} \sin \omega t,$$

where

$$E_{cm} = 2E_m \sin\frac{\pi}{p}. \quad (2.26)$$

The relationship between E_m, U_m, and E_{cm} is illustrated by the phasor diagrams drawn in Figure 2.3 for $p = 2, 3, 4$, and 5. The supply system voltages are shown by light-line phasors of length E_m, a possible peak thyristor voltage by the heavy line of length U_m, and one of the commutating voltages by the dashed-line phasor of length E_{cm}. For $p = 2$ and $p = 3$, note that $U_m = E_{cm}$.

At the time of commutation, the value E_c of the commutating voltage e_c is

$$E_c = E_{cm} \sin \alpha.$$

Therefore, at the thyristor current zeros, if we assume $I_d > I_s$,

$$\frac{di}{dt} = \frac{E_c}{(2L_c + L_u)},$$

which is a maximum when the firing angle α is $\pi/2$,

$$\left(\frac{di}{dt}\right)_{\max} = \frac{E_{cm}}{(2L_c + L_u)}.$$

If nonlinear reactors are not used, $L_u = 0$ and

$$\left(\frac{di}{dt}\right)_{\max} = \frac{E_m \sin(\pi/p)}{L_c}. \tag{2.27}$$

The forward dv/dt applied to the thyristors after commutation is

$$\frac{dv}{dt} = -\frac{de_c}{dt} = -\omega E_{cm} \cos \omega t.$$

This expression is positive in the inverter region of operation and is a maximum at the time when the commutating voltage crosses zero, going positive ($\omega t = \pi$),

$$\left(\frac{dv}{dt}\right)_{\max} = \omega E_{cm} = 4\pi f E_m \sin\frac{\pi}{p}. \tag{2.28}$$

Note that at other times during the cycle, the instantaneous dv/dt can be greater by the ratio U_m/E_{cm}, and notches caused by other commutations may produce theoretically infinite dv/dt, requiring suppression by snubbers. (See Chapter 8.)

The peak value of thyristor current is the assumed constant dc current,

$$J_m = I_d. \tag{2.29}$$

When overlap is neglected, the conduction angle of each thyristor is

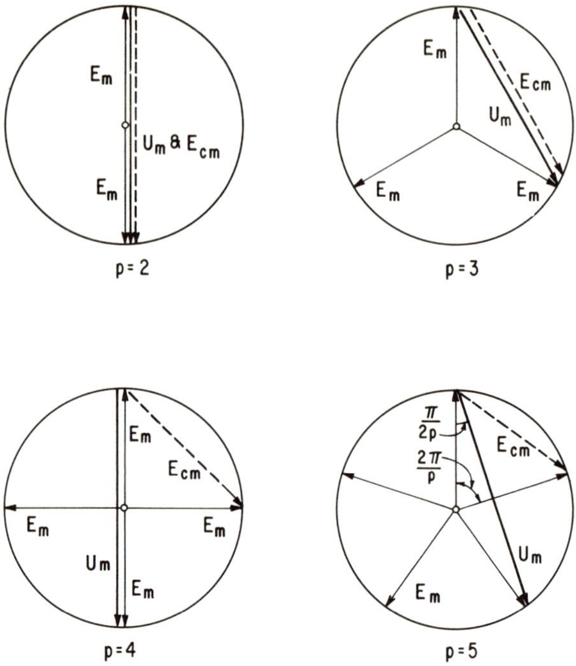

Figure 2.3. Phasor diagrams showing relationship between supply system voltages, peak possible thyristor voltage, and commutating voltage.

$2\pi/p$, or the duty cycle is $1/p$. Hence, the average and rms thyristor currents are, respectively,

$$J_0 = \frac{I_d}{p}; \tag{2.30}$$

$$J_{e0} = \frac{I_d}{\sqrt{p}}. \tag{2.31}$$

When overlap is considered, the average current remains the same, because the current "lost" during the rise time is equal to the current "gained" during the fall time. However, a correction (normally quite small, see Schaefer,[7] p. 334) must be applied to find the true rms current,

$$J_e = J_{e0}\sqrt{1 - p\psi(\alpha,u)},$$

where the function $\psi(\alpha,u)$ is defined as

$$\psi(\alpha,u) = \frac{\sin u\,[2 + \cos(2\alpha + u)] - u[1 + 2\cos\alpha\cos(\alpha + u)]}{2\pi\,[\cos\alpha - \cos(\alpha + u)]^2}.$$

2.4 INPUT CURRENT AND POWER FACTORS

The rms line current of a simple rectifier is

$$I_e = J_e\sqrt{n_w} = I_d\sqrt{\frac{n_w}{p}}\;\sqrt{1 - p\psi(\alpha, u)}.$$

A single-way connection ($n_w = 1$) has a dc component of current. If a transformer is used, this component (J_0) is eliminated from the primary winding. The primary current referred to the secondary side is then

$$I'_e = \sqrt{J_e^2 - J_0^2}.$$

The crest value of the fundamental component of line current, neglecting overlap, is

$$I_{1m0} = I_d\left(\frac{2n_w}{\pi}\right)\sin\left(\frac{\pi}{p}\right) \tag{2.32}$$

and lags the phase voltage by the firing delay angle α. The displacement power factor is, therefore, $\cos\alpha$. If one specifies the phase voltage as

$$e = E_m\cos\omega t, \tag{2.33}$$

the current may be resolved into real and reactive components,

$$\begin{aligned} i_1 &= I_{1m0}\cos(\omega t - \alpha) \\ &= I_{1m0}\cos\alpha\cos\omega t + I_{1m0}\sin\alpha\sin\omega t. \end{aligned} \tag{2.34}$$

When overlap is considered, it can be shown that Equation 2.34 becomes

$$\begin{aligned} i_1 &= I_{1m0}\,p_r\cos\omega t + I_{1m0}\,p_x\sin\omega t \\ &= I_{1m0}\sqrt{p_r^2 + p_x^2}\cos(\omega t - \phi); \end{aligned}$$

the displacement power factor is

$$\cos\phi = \frac{p_r}{\sqrt{p_r^2 + p_x^2}} \tag{2.35}$$

where

$$p_r = \frac{[\cos\alpha + \cos(\alpha + u)]}{2} \tag{2.36}$$

$$p_r = E_d/E_{do} \qquad \text{(neglecting losses, see Equation 2.12)}$$

and

$$p_x = \frac{2u + \sin 2\alpha - \sin 2(\alpha + u)}{4\left[\cos \alpha - \cos(\alpha + u)\right]} \; . \tag{2.37}$$

Note that Equations 2.32 to 2.37, without the factor n_w in Equation 2.32, also give the fundamental components of current in a thyristor.

The total power factor P_t of the windings supplying a component simple rectifier, or the inverse of the utilization factor of those windings, is given by

$$P_t = \frac{E_d I_d n_w}{E_e I_e p s} \, .$$

This may be written, neglecting losses, in terms of the factors $\psi(\alpha,u)$ and p_r, which are directly related to α and u,

$$P_{to} = \frac{p_r \sqrt{2pn_w}\, \sin(\pi/p)}{\pi\sqrt{1 - p\psi(\alpha,u)}} \, .$$

The complete normalized Fourier series for the current in a thyristor is, with time t measured from the crest of the phase voltage as in Equation 2.33,

$$\begin{aligned}\frac{i}{I_d} = {} & \frac{1}{p} + \frac{2}{\pi} \sin\frac{\pi}{p}\left[\frac{\cos\alpha + \cos(\alpha + u)}{2}\right] \cos \omega t \\ & + \frac{2}{\pi} \sin\frac{\pi}{p}\left[\frac{2u + \sin 2\alpha - \sin 2(\alpha + u)}{4\{\cos\alpha - \cos(\alpha + u)\}}\right] \sin\omega t \\ & + \sum_{n=2}^{\infty} \left[b_n \cos n\omega t + a_n \sin n\omega t\right]\end{aligned}$$

where

$$\begin{aligned} a_n = K_n & \left[\frac{\sin(n-1)(\alpha + u) - \sin(n-1)\alpha}{n-1}\right. \\ & \left. - \frac{\sin(n+1)(\alpha + u) - \sin(n+1)\alpha}{n+1}\right] \end{aligned}$$

$$\begin{aligned} b_n = K_n & \left[\frac{\cos(n-1)(\alpha + u) - \cos(n-1)\alpha}{n-1}\right. \\ & \left. - \frac{\cos(n+1)(\alpha + u) - \cos(n+1)\alpha}{n+1}\right] \end{aligned}$$

$$K_n = \frac{\sin(n\pi/p)}{n\pi[\cos\alpha - \cos(\alpha + u)]} \,. \tag{2.38}$$

Note that these equations become indeterminate (0/0) when $u = 0$, in which case, for $n = 1$ to ∞,

$$a_n = \frac{2}{n\pi} \sin\left(\frac{n\pi}{p}\right) \sin n\alpha,$$

$$b_n = \frac{2}{n\pi} \cos\left(\frac{n\pi}{p}\right) \cos n\alpha.$$

The resultant crest magnitude of nth harmonic component is

$$c_n = \sqrt{a_n^2 + b_n^2}\,.$$

For the line current in double-way circuits, the fundamental and odd harmonics are doubled while the dc and even harmonics cancel.

3 Current Surges in Converter Circuits

As in any type of electrical power apparatus, abnormally high current surges can occur in some components of converter equipment as a consequence of faults or failures. Since semiconductors have a very small overload capacity compared with other power devices, it is particularly important to know what magnitude and duration of surge currents can be expected and what might be done to limit them. In power converters, surges can be caused by faults beyond the output terminals, faults before the input terminals, and malfunction or failures within the equipment itself. However, all significant faults usually result in the breakdown or conduction of a semiconductor device that should be in the blocking state.

The transition of a thyristor from the blocking state into conduction, whether by catastrophic failure or by misfiring, is abrupt in the manner of a switch. Hence, it is natural that the waveforms of current surges should be similar to ideal switching transients. In this section, characteristic data for a worst-case switching transient with an ac source and an inductive impedance are presented, followed by application of the results to surge currents in phase-controlled converters. Various types of converter faults and protective measures are discussed, with emphasis on preventing damage to thyristors in the equipment.

3.1 GENERAL ANALYSIS OF SURGE CURRENTS

The surge conditions produced by faults in converter circuits are analytically equivalent to the switching transients in an inductive circuit driven by a sinusoidal voltage. The worst-case condition having maximum transient offset occurs when the switch is closed (or the fault begins) at a voltage zero crossing, or time $t = 0$ for a voltage $E_m.\sin \omega t$. With pure inductance, the current waveform shown in Figure 3.1(A) is obtained. However, the relatively small amount of resistance present in practical circuits is sufficient to appreciably reduce the surge current, indicated by the shaded area in Figure 3.1(B) as the difference between a steady-state component and the negative of a transient component.

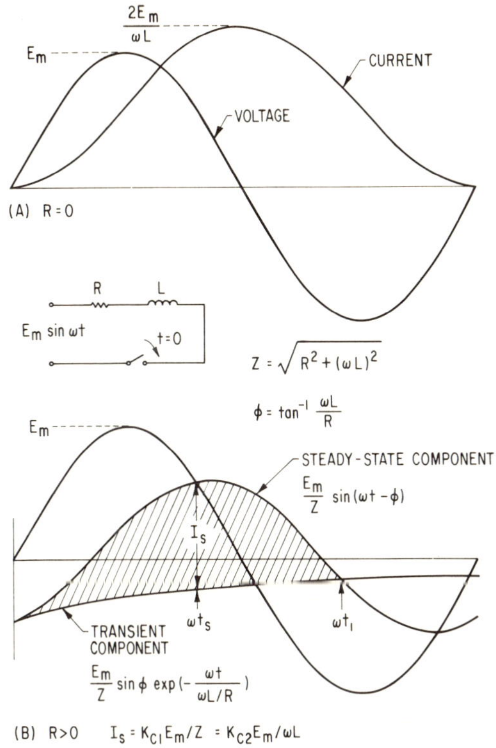

Figure 3.1. Worst-case switching or surge transients.

$$i = \frac{E_m}{Z}\left[\sin(\omega t - \phi) + \sin\phi \exp -\left(\frac{\omega t}{\omega L/R}\right)\right] \tag{3.1}$$

where

$$Z = \sqrt{R^2 + (\omega L)^2}$$

and

$$\phi = \mathrm{Tan}^{-1}(\omega L/R).$$

Note that the quantities used in this section are separately defined and should not be confused with the specific converter circuit nomenclature used elsewhere.

The peak or crest value of surge current I_s can be related to the peak value of the steady-state component E_m/Z by a crest factor K_{c1}, or by a factor K_{c2} to the ideal peak steady-state current $E_m/\omega L$ that would be obtained if R were negligible,

$$I_s = K_{c1}\frac{E_m}{Z} = K_{c2}\frac{E_m}{\omega L}\,. \tag{3.2}$$

Both of these factors are a function of the ratio $\omega L/R$, and their values for $\omega L/R$ from 1 to 10 are listed in Table 3.1, together with the angle ωt_s where the peak I_s occurs and the angle ωt_1 where the surge returns to zero.

The reduction in peak surge current obtained with a relatively small resistance can be seen from the table. For example, a resistance equal to 20% of the reactance ($\omega L/R = 5$) reduces the peak current by the ratio 1.520/2.0 or to 76% of the value that would be predicted if the resistance were neglected.

The surge capability of thyristors and the performance of current limiting fuses used for their protection are often expressed in terms of I^2t, which is a measure of the heat generated by resistance losses. This is a useful parameter, although it is not a constant quantity if the surge pulse is long enough for some heat to escape. The total I^2t of the shaded current pulse in Figure 3.1 can be calculated by squaring and integrating Equation 3.1 over the range 0 to t_1. Computed values of the normalized quantity $\omega\int_0^{t_1} i^2dt/(E_m/Z)^2$ are listed in the last column of Table 3.1. It is seen that a relatively small resistance reduces I^2t even more than the peak current.

Table 3.1. Surge Current Factors

$\omega L/R$	K_{c1}	K_{c2}	ωt_s (degrees)	ωt_1 (degrees)	I^2t (per unit)
0	1.000	0	90.0	180.0	1.253
1	1.069	0.756	130.9	225.8	1.977
2	1.242	1.111	146.3	249.3	2.877
3	1.379	1.308	154.2	263.4	3.702
4	1.477	1.433	159.0	273.1	4.376
5	1.550	1.520	162.3	280.3	4.922
6	1.606	1.584	164.7	286.0	5.367
7	1.649	1.633	166.5	290.5	5.736
8	1.684	1.671	167.9	294.3	6.045
9	1.713	1.702	169.1	297.6	6.307
10	1.737	1.728	170.0	300.4	6.533
∞	2.000	2.000	180.0	360.0	9.425

3.2 CAUSES AND LIMITATION OF THYRISTOR SURGE CURRENTS

3.2.1 Thyristor Failure to Block Reverse Voltage

This type of fault occurs in diode rectifiers as well as controlled rectifiers, inverters, and cycloconverters. Typical conditions are shown by Schaefer,[7] pp. 181–182. A pair of input phases are short-circuited and the surge of fault current is limited by the commutating impedance. In diode rectifiers, fuses are the prime protective devices. With thyristors, a fault detector can be used to block the firing signals, limiting the fault to one cycle. Fuses are used as backup protection, and in high-power equipment using multiple devices in each path, to limit the current in the failed device to below the explosion level and remove that device from the circuit. In the latter case, it may then be possible to unblock the thyristors and resume operation. The same is true for mercury-arc equipment where the fault is likely an arc back from which recovery is possible. However, in single-path thyristor equipment, the blocking should be permanent and backed up by tripping the supply breaker.

If one assumes that only two lines are involved in the fault (as when the other thyristors are successfully blocked), the worst-case surge is driven by one cycle of the voltage between a pair of lines having the greatest voltage difference. An example of the surge current path is shown in Figure 3.2. If one uses the crest factor K_{c2} defined in Section 3.1, the peak current, which should be less than the 1-cycle nonrepetitive surge rating of the thyristors so that the non-failed device will not be damaged, is given by

$$J_{s1} = K_{c2} \frac{U_m}{(2\omega L_c)}. \tag{3.3}$$

When two or more thyristors are connected in series with a voltage safety factor greater than 2, then failure of one device should not disturb the converter. An indicator of such failure should be provided so that the device can be replaced during a normal shutdown. The incidence of double failures should be negligible, so that this type of fault can be ignored as a criterion for determining thyristor rating.

3.2.2 Intergroup Faults

In cycloconverters or reversible dc converters, a pair of supply lines may become short circuited through two forward-conducting thyristors, one in each of the opposite current-polarity groups. This may be caused by a failure of the blanking logic, misfiring due to noise or

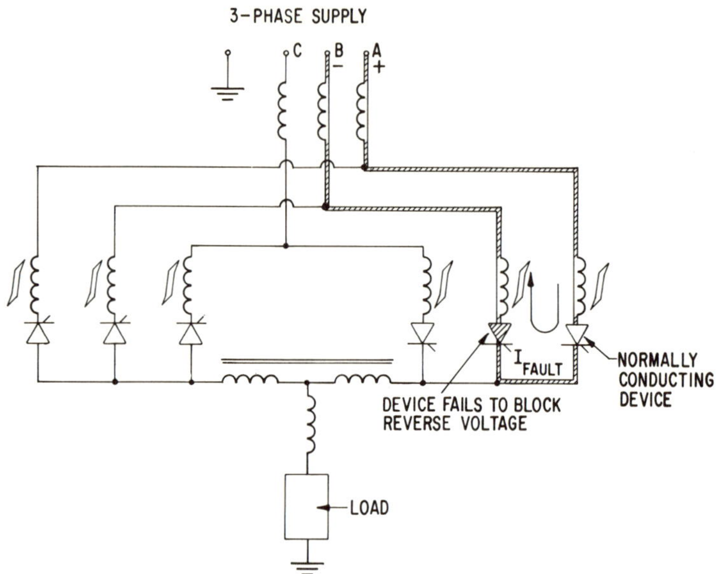

Figure 3.2. 3-Phase single-way cycloconverter, showing fault caused by device failure. The path of the surge or fault current is indicated by the heavy line. It is assumed that the voltage of phase A is instantaneously positive with respect to phase B.

other gate pulse generator trouble, or forward breakdown of a thyristor. An example of the surge current path is shown in Figure 3.3, and the worst-case waveform is similar to Figure 3.1. If no intergroup reactors are used, the peak surge current is again given by Equation 3.3. With intergroup reactors, the current is reduced to

$$J_{sr} = K_{c2} \frac{U_m}{2\omega \, (L_c + L_g + M_g)} . \tag{3.4}$$

Note that the current path in Figure 3.3 is the same as that of the circulating currents that flow when blanking is not used and both groups of thyristors receive firing signals at all times. Of course, the firing angles are such that the worst-case condition shown in Figure 3.1 is never attained in normal operation. Intergroup reactors are usually employed to limit these circulating currents. Even if circulating currents are normally eliminated by blanking logic or offset bias signals, it may still be desirable to provide intergroup reactors to reduce fault currents.

Since intergroup faults can be caused by other than catastrophic

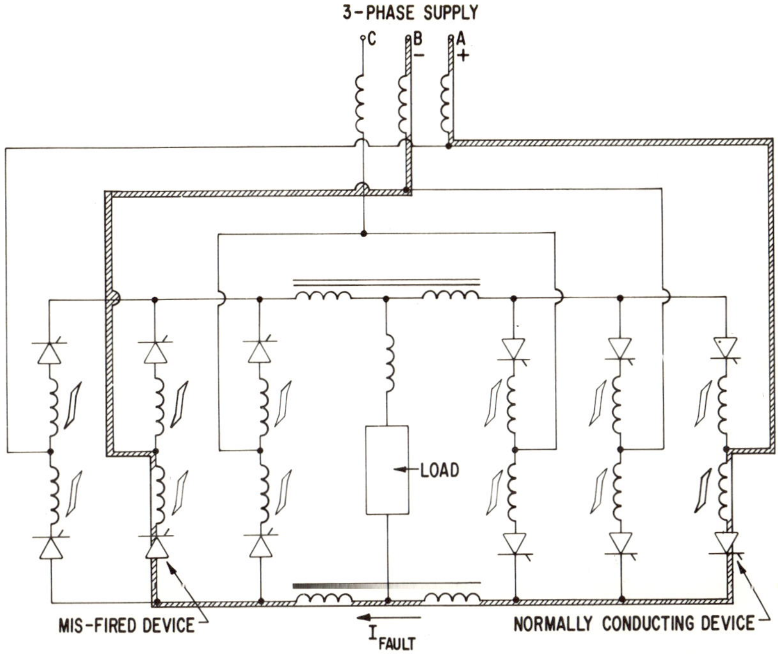

Figure 3.3. 3-Phase double-way cycloconverter, showing intergroup fault caused by misfire of a device. The path of the surge or fault current is indicated by the heavy line. It is assumed that the voltage of phase A is instantaneously positive with respect to phase B.

failure of components, recovery may be possible if no corrective action or temporary corrective action is taken in response to detection of such faults. Permanent blocking of the thyristors may not be necessary. However, the faults are liable to be repetitive so that the current J_{sr} should be limited to some repetitive surge capability of the thyristors. Besides performing such a limiting function, intergroup reactors will also assist in filtering the output and aid the protective logic circuitry in discriminating among types of faults.

3.2.3 Commutation Failure

Commutation failures occur in cycloconverters and other phase-controlled circuits during the inverter mode of operation when the margin angle γ is lost. They may be caused by reduction of the supply voltage, including commutation notches and faults in other converters, overload currents, and various control circuit malfunctions. An example is given by Schaefer,[7] pp. 187–189. Corrective action commonly consists of advancing the firing angle to regain the margin. In dc motor drives, dc breakers may be necessary. For cycloconverters, the thyristors may be blocked and the load current will be extinguished after the counter emf reverses.

The fault current through the thyristors may be severe but is so dependent upon the type of load and operating conditions that no generally applicable equation can be written. However, it should be possible to design a protective scheme such that a commutation failure is no more severe than other types of faults in determining the thyristor rating required.

3.2.4 Load Faults

The control circuits should be provided with an overcurrent detector in a fast-acting feedback control loop that will take over from the normal phase-control circuitry and limit the load current to well within the capability of the thyristors during normal overload conditions and moderate faults. However, in severe faults such as a short circuit across the load terminals, the rise of current may be so fast that it cannot be arrested by phase control, particularly if filter inductors are not used or are very small, as in the case of cycloconverters that must operate with a high frequency ratio.

In this case, the current will be limited by the commutating inductance and any intergroup inductance and will be about the same order of magnitude as Equation 3.3 or 3.4. Polyphase converters depart from regular or mode 1 operation and follow a characteristic dependent upon the distribution of inductance in the circuit and the nature of the thyristor firing signals. By appropriate control of the thyristors, the load fault current can be reduced and extinguished.

4 The Forward-Reverse Cycloconverter with Low-Frequency Output

The term "forward-reverse" type is used to distinguish cycloconverters that have a configuration similar to reversing dc motor drives (for each phase of output) from the "delta-rectifier" type of cycloconverter described in Chapter 9. Separate groups of thyristors are used to provide positive or "forward" load current and negative or "reverse" load current. This is presently the most common type of cycloconverter.

In this chapter, operation with an output frequency very much lower than the supply frequency will be considered. The forward-reverse cycloconverter will be treated as a reversing dc converter with sinusoidal modulation. The effects of higher frequency and nonsinusoidal modulation are analyzed in Chapters 5 and 6, respectively.

First, the basic equations for modulation of the firing delay angles in order to achieve a sinusoidal output voltage are derived. Then, a sinusoidal load current is postulated and the current-carrying duty of the thyristors is discussed in Section 4.2. The current in the supply lines is considered in Section 4.3, leading to expressions for the modulation of real and reactive power drawn from the source and computation of the average displacement power factor of a cycloconverter.

In Section 4.4, it is shown how the equivalent circuit of a phase-controlled rectifier can be adapted to suit the case of a cycloconverter, and how it can be modified to include the correction factors for high output/input frequency ratio to be derived in Chapter 5. Finally, in Section 4.5, the relationship between the voltage regulation of the converter and the *di*/*dt* and surge current capability of the thyristors is investigated. These factors are all dependent on the commutating inductance and become quite important when the source frequency is high.

4.1 SINUSOIDAL OPERATION, NEGLECTING OVERLAP

When the phase-control signal is modulated, the average dc voltage and current variables in the rectifier equations presented in Section 2.1 can be regarded as "pseudo-instantaneous" quantities. The output voltage and current will change little during the conduction period of a thyristor ($1/fp$ sec) if the modulating or output frequency F is small compared with the supply frequency f. As in the case of the rectifier, the firing angle α and the load current are taken as the independent variables.

If the desired output voltage is sinusoidal,

$$v_0 = V_{m0} \sin \Omega t, \tag{4.1}$$

where v_0 is the pseudo-instantaneous voltage neglecting overlap, corresponding to $E_{d\alpha}$ in Equation 2.2 The firing angles α_p of the positive current group and α_n of the negative current group must be modulated to satisfy the relationship

$$E_{d0} \cos \alpha_p = - E_{d0} \cos \alpha_n = V_{m0} \sin \Omega t$$

or

$$\cos \alpha_p = k \sin \Omega t, \tag{4.2}$$

where k is the modulation factor, defined as

$$k = V_{m0}/E_{d0}, \qquad 0 < k < 1. \tag{4.3}$$

Hence

$$\alpha_p = \operatorname{Cos}^{-1} (k \sin \Omega t), \qquad 0 < \alpha_p < \pi \tag{4.4}$$

and

$$\alpha_n = \pi - \alpha_p. \tag{4.5}$$

Capitalization of the inverse trigonometric function denotes the principal value, which is restricted to the range indicated.

The firing angle modulation of the two polarity groups may alternatively be described in terms of an angle ψ, which varies about the zero voltage point ($\alpha = \pi/2$) and is defined as

$$\psi = \frac{\pi}{2} - \alpha_p = \alpha_n - \frac{\pi}{2} \tag{4.6}$$

and is restricted to the range

$$-\frac{\pi}{2} < \psi < \frac{\pi}{2}.$$

Thus

$$\sin \psi = \cos \alpha_p = - \cos \alpha_n \tag{4.7}$$

or

$$\sin \psi = k \sin \Omega t$$

$$\psi = \operatorname{Sin}^{-1} (k \sin \Omega t).$$

Note that by definition Equation 4.6, the ψ is the same for both positive and negative groups, unlike α. The basic equation for the output voltage in terms of ψ is

$$v_0 = E_{d0} \sin \psi. \tag{4.8}$$

4.2 THYRISTOR CURRENT DUTY

The load current is assumed to be sinusoidal, the components at harmonics of the supply frequency being suppressed by the filtering action of the load inductance, aided perhaps by intergroup inductance L_g or a filter inductance L_f added for this purpose. The current is assumed to lag the programmed voltage wave v_0, Equation 4.1, by a phase angle Φ,

$$a = A_m \sin(\Omega t - \Phi). \tag{4.9}$$

The current supplied by the positive group is shown in Figure 4.1(B), and the current in a particular thyristor of this group has the waveform in Figure 4.1 (C), when intergroup circulating currents are suppressed. From this figure, the peak, average, and rms thyristor currents are seen to be, respectively,

$$\begin{aligned} J_m &= A_m; \\ J_0 &= \frac{A_m}{\pi p}; \\ J_{e0} &= \frac{A_m}{2\sqrt{p}}. \end{aligned} \tag{4.10}$$

Since each line supplies thyristors in both polarity groups and a second such pair in the case of double-way circuits, the rms line current is

$$I_{e0} = J_{e0}\sqrt{2n_w} = A\sqrt{\frac{n_w}{p}}.$$

When overlap is considered, the rms value is slightly reduced, but Equation 4.10 gives a safe value for design purposes. Note that the rise and fall di/dt in the thyristors during the overlap period is proportional to sin α, as shown in Figure 4.1(A) and (D). Thus, the worst-case switching duty of the thyristors will occur when $\Phi = \pm\pi/2$ (purely reactive load), where the peak current A_m is switched at the maximum rate, given by Equation 2.27. When the output voltage is reduced by decreasing the modulation of α about the median $\pi/2$, the cyclic variation of di/dt becomes less and its average value is greater.

If the output frequency is so low that the load current dwells in the vicinity of its peak value A_m for a period comparable with the thermal time constant of the thyristors and their cooling system, then the

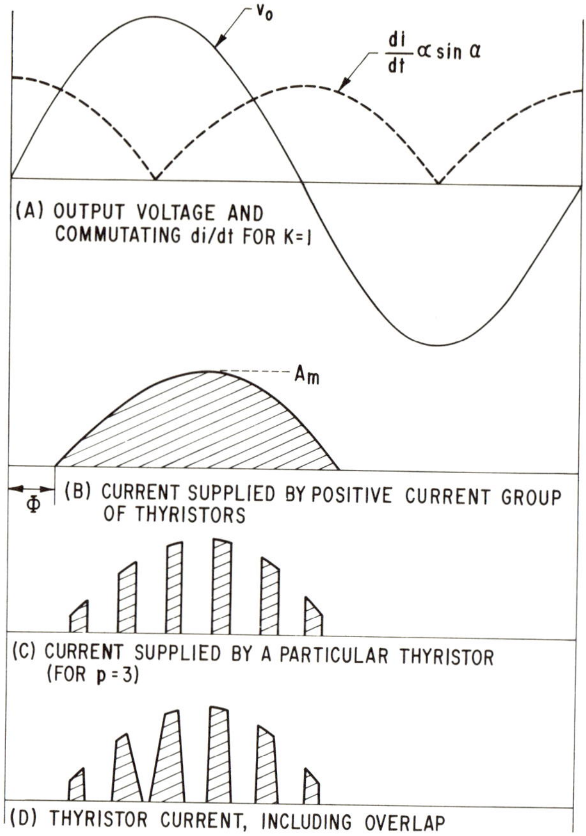

Figure4.1. Current waveforms in a cycloconverter.

thyristors must effectively be sized to carry the peak current continuously. The rating of the thyristors must be determined from Equations 2.29 to 2.31 with $I_d = A_m$. The required average current rating is increased by a factor π and the rms current rating is doubled. Since presently published thyristor ratings, valid over the range 50 to 400 Hz, are in part determined by the peak value of junction temperature fluctuations at 50 Hz, it is expected that some derating must be applied for any lower output frequency.

An effective increase in the thyristor current rating is also required when the output frequency is equal to, or a low subharmonic of, the supply frequency, such as 15, 30, or 60 Hz output with a 60 Hz supply. The worst-case duty for a given thyristor will occur at a particular phase relationship between the load and source voltage waveforms and for a certain current phase Φ. This thyristor will be repetitively subjected to more than its share of load current.[86] Its conduction period may be extended for longer than the average time $1/fp$ (see Chapter 5 on the effect of frequency ratio) and probably coincides with the peak load current. For other phase relationships, the load will be shared more uniformly among the thyristors in the group, or another thyristor in the group will be subjected to the same worst-case duty. If the output frequency is not locked but is close to such a critical harmonic relationship with the source, then the incidence of worst-case duty will "beat" cyclically from one thyristor to the next. Although the load will be shared uniformly on a long-time average basis, the beat frequency may be so low that the worst-case condition with locked frequencies must be applied in determining the necessary size of thyristors.

4.3 LOADING OF THE SUPPLY

The analysis in this section follows the results of von Issendorff[77] in calculating the reactive power and average displacement power factor of a cycloconverter.

If at time $t = 0$, it is assumed that phase 1 of the source is displaced from the programmed output voltage defined in Equation 4.1 by a phase angle λ (at the source frequency f), the supply voltages may be written as

$$e_i = E_m \sin\left[\omega t + \lambda - (i-1)\frac{2\pi}{p}\right] \qquad \text{for } i = 1 \text{ to } p.$$

In particular, the voltage of phase 1 is

$$e_1 = E_m \sin(\omega t + \lambda). \tag{4.11}$$

The crest value of the fundamental component of line current is modulated at twice the output frequency and is given by Equation 2.32 with $|u|$ from Equation 4.9 substituted for I_d:

$$\begin{aligned} I_{1m0} &= A_m |\sin(\Omega t - \Phi)| \left(\frac{2n_w}{\pi}\right) \sin\left(\frac{\pi}{p}\right) \\ &= A'_m |\sin(\Omega t - \Phi)|, \end{aligned} \tag{4.12}$$

where A'_m is defined to be

$$A'_m = A_m \left(\frac{2n_w}{\pi}\right) \sin\left(\frac{\pi}{p}\right). \tag{4.13}$$

When the positive group is conducting the load current, the supply current lags by a phase angle α_p (neglecting overlap). Similarly, the phase lag is α_n when the negative group conducts. Hence, if one uses Equation 4.6, the instantaneous value i_1 of the fundamental component of current in supply phase 1 may be written:

$$i_1 = I_{1mo} \sin(\omega t + \lambda - \alpha_p) = I_{1mo} \sin(\omega t + \lambda - \frac{\pi}{2} + \psi) \text{ for } a > 0; \tag{4.14}$$

$$i_1 = I_{1mo} \sin(\omega t + \lambda - \alpha_n) = I_{1mo} \sin(\omega t + \lambda - \frac{\pi}{2} - \psi) \text{ for } a < 0. \tag{4.15}$$

If one uses the sign function (as employed in the BASIC programming language) defined by the following table:

	$x < 0$	$x = 0$	$x > 0$
$\text{sgn}(x)$	-1	0	$+1$

,

Equations 4.14 and 4.15 can be written as one,

$$\begin{aligned} i_1 &= I_{1mo} \sin\left[\omega t + \lambda - \frac{\pi}{2} + \psi \text{ sgn}(a)\right] \\ &= A'_m |\sin(\Omega t - \Phi)| \sin\left[\omega t + \lambda - \frac{\pi}{2} + \text{Sin}^{-1}(k \sin\Omega t) \text{ sgn}(\Omega t - \Phi)\right]. \end{aligned} \tag{4.16}$$

The product of Equations 4.11 and 4.16 gives the instantaneous fundamental power drawn from phase 1. The phase difference ϕ between the voltage and current is seen to be

$$\phi = \frac{\pi}{2} - \text{sgn}(\Omega t - \Phi) \text{ Sin}^{-1}(k \sin \Omega t).$$

The pseudo-instantaneous real power (i.e., the algebraic average power over one cycle of the supply frequency) is given by

$$\begin{aligned} P_r &= 0.5 E_m I_{1mo} \cos\phi \\ &= 0.5 E_m A'_m \, |\sin(\Omega t - \Phi)| \, \sin[\text{sgn}(\Omega t - \Phi) \text{ Sin}^{-1}(k \sin \Omega t)] \\ &= 0.5 E_m A'_m \sin(\Omega t - \Phi) \, k \sin \Omega t, \end{aligned} \tag{4.17}$$

because

$$\sin(\pm x) = \pm \sin x.$$

The average real power P_R, or the algebraic average of P_r over one cycle of the output frequency, is easily seen to be

$$P_R = 0.25E_m A'_m k \cos \Phi. \tag{4.18}$$

With the defining Equations 2.1, 4.3, and 4.13, the input-output power balance is verified,

$$\frac{P_R\, ps}{n_w} = 0.5V_{mo} A_m \cos \Phi.$$

Similarly, the pseudo-instantaneous reactive power P_x is defined as

$$\begin{aligned} P_x &= 0.5E_m I_{1mo} \sin \phi \\ &= 0.5E_m A'_m |\sin(\Omega t - \Phi)| \cos[\mathrm{sgn}(\Omega t - \Phi)\ \mathrm{Sin}^{-1}(k \sin \Omega t)] \\ &= 0.5E_m A'_m |\sin(\Omega t - \Phi)| \sqrt{1 - k^2 \sin^2 \Omega t}. \end{aligned} \tag{4.19}$$

Note that, because

$$\cos(\pm x) = \cos x,$$

the sign function is lost, and only the positive square root, denoted by the radical sign, is applicable. Thus, Equation 4.19 expresses mathematically the well-known fact that the reactive power drawn from the source is always positive or that cycloconverters impose a lagging load on the supply system for all conditions of output loading.

In Figure 4.2(A), the parameters $k \sin \Omega t$, $\sqrt{1 - k^2 \sin^2 \Omega t}$, $\sin(\Omega t - \Phi)$, and $|\sin(\Omega t - \Phi)|$ are sketched over one cycle of the output frequency to illustrate the derivation of the pseudo-instantaneous real and reactive power quantities P_r and P_x that are shown in Figure 4.2(B). The light chain lines in Figure 4.2(A) indicate the limit values of the term $\sqrt{1 - k^2 \sin^2 \Omega t}$ for the extremes $k = 0$ and $k = 1$. Special cases of the curves in Figure 4.2 are shown in Figure 4.3 for $k = 1$, $\Phi = 0$; in Figure 4.4 for $k = 1$, $\Phi = \pi/2$; in Figure 4.5 for $k = 0$, $\Phi = \pi/3$. In the latter case, the curves for other values of Φ are simply phase displaced.

The average reactive power P_X can be obtained by integrating Equation 4.19 over a half-cycle interval of the output, which for convenience can be taken as the conduction time of the positive current group of thyristors.

$$P_X = \frac{E_m A'_m}{2\pi} \int_{\Phi}^{\Phi+\pi} |\sin(\Omega t - \Phi)| \sqrt{1 - k^2 \sin^2 \Omega t}\ d(\Omega t). \tag{4.20}$$

Then, the average displacement power factor of the converter is

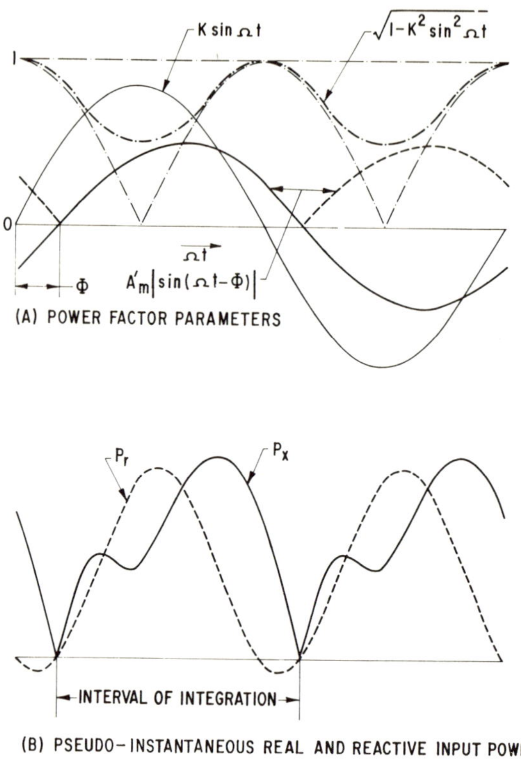

Figure 4.2. Input power of forward-reverse cycloconverter with sinusoidal output.

$$P_1 = \frac{P_R}{\sqrt{P_R^2 + P_X^2}} = \frac{1}{\sqrt{1 + (P_X/P_R)^2}}; \tag{4.21}$$

and

$$\frac{P_X}{P_R} = \frac{\int_{\Phi}^{\Phi+\pi} \sin(x - \Phi)\sqrt{1 - k^2 \sin^2 x}\, dx}{0.5\pi\, k \cos\Phi}. \tag{4.22}$$

In the general case, the integral term involves elliptic functions and therefore is best evaluated by numerical integration using a computer. The special case of unity modulation ($k = 1$) reduces to

$$\int_{\Phi}^{\Phi+\pi} \sin(x - \Phi)\,|\cos x|\, dx = \Phi \sin\Phi + \cos\Phi;$$

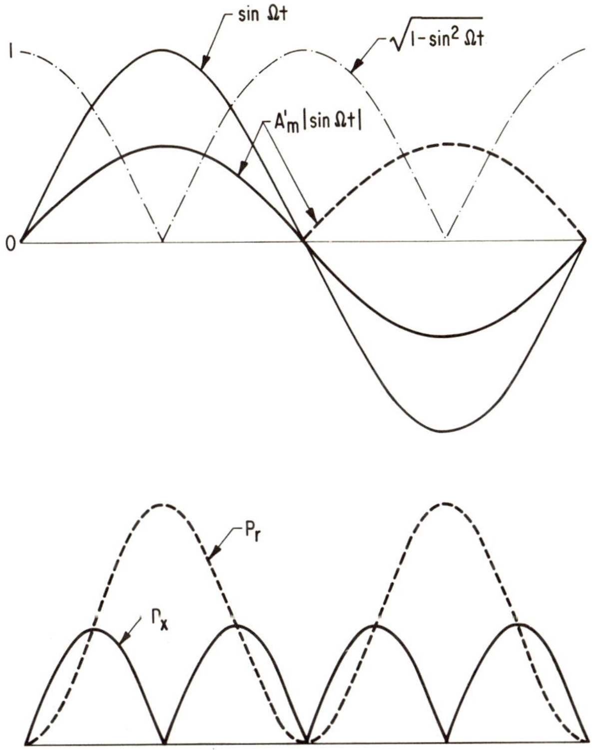

Figure 4.3. Special case of Figure 4.2 for full modulation and resistive load. ($k = 1$, $\Phi = 0$).

and
$$\frac{P_X}{P_R} = \frac{2}{\pi}(\Phi \tan \Phi + 1).$$

An approximate solution can be obtained for the case of low modulation ($k \ll 1$) by taking the first two terms in the series expansion of the square-root term

$$\int_{\Phi}^{\Phi+\pi} \sin(x - \Phi)(1 - 0.5k^2 \sin^2 x)\, dx = 2 - \frac{k^2(1 + \cos^2 \Phi)}{3}. \quad (4.23)$$

The displacement power factor P_1 obtained by numerical integration of Equation 4.22 is plotted in Figure 4.6. The values calculated by using approximate Equation 4.23 differ from the true values by less than 0.005 for $k \leqslant 0.6$.

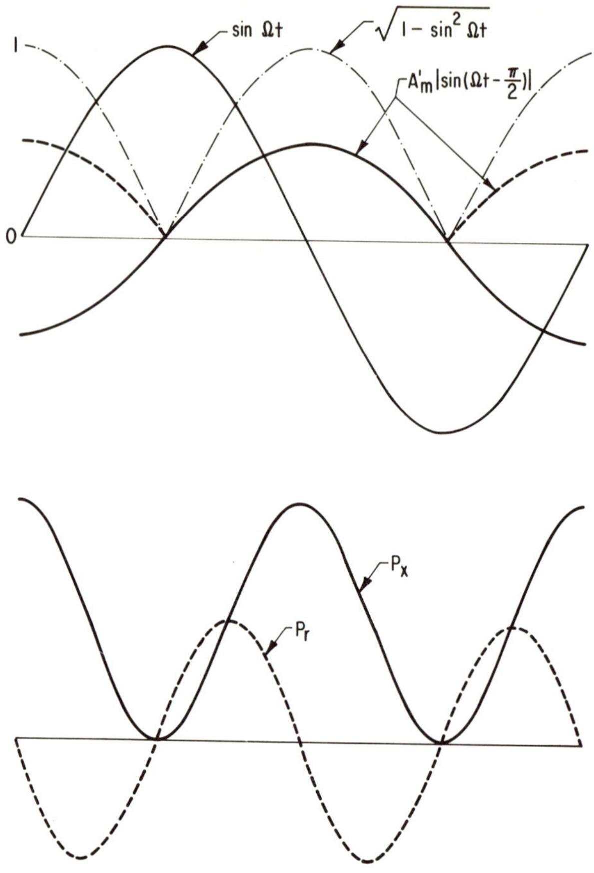

Figure 4.4. Special case of Figure 4.2 for full modulation and reactive load. ($k = 1$, $\Phi = \pi/2$).

A maximum input power factor of 0.844 is attained when the output power factor and the modulation factor are both unity. When either of these factors are less than unity, the input power factor is also reduced. For example, with $k = 0.8$ and $\cos \Phi = 0.8$, then $\cos \phi = 0.535$, which is approximately equal to $0.8 \times 0.8 \times 0.844$. These values are for theoretically ideal conditions and do not include the distortion power factor due to harmonic currents. For regenerative loads, both cos Φ and cos ϕ are negative, but Figure 4.6 is still valid for relating their magnitudes. Remember again that a cycloconverter always draws lagging reactive current from its supply system, i.e., $0 \leqslant \phi \leqslant \pi$, while Φ may vary over the full range $-\pi \leqslant \Phi \leqslant \pi$.

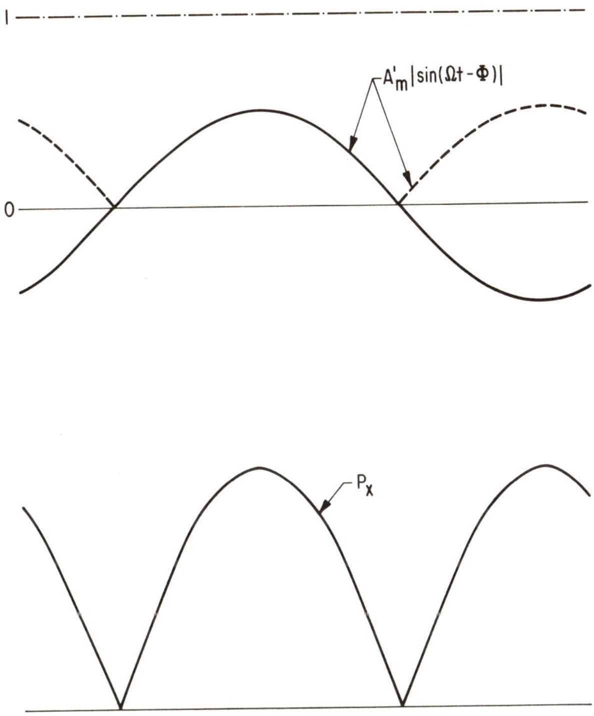

Figure 4.5. Special case of Figure 4.2 for zero modulation. ($k = 0$, $\Phi = \pi/3$).

Methods of improving the input power factor are discussed in Chapter 6.

With polyphase output provided by n_0 converters operating from the same supply system, the total pseudo-instantaneous real (or reactive) power from each input line is the sum of n_0 phase-displaced curves of the type shown in Figures 4.2(B) to 4.5(B). For balanced loading, the total real power of a simple converter is always constant at the value n_0pP_R, of course. The total reactive power has an average value n_0pP_X, but it fluctuates in a manner that depends upon both k and Φ. When $k = 1$ and $\Phi = 90°$ (see Figure 4.4), the total reactive power of n_0 phases is steady. For $k = 1$ and $\Phi = 0$, the reactive power curve shown in Figure 4.3, when summed with similar but phase-displaced curves, produces a fluctuation at a frequency $4F\,n_0$ having a shape similar to the voltage ripple of an n_0-phase diode rectifier. If $n_0 = 3$, the ratio of minimum to maximum reactive power is $\sqrt{3}/2$. If $k = 0$, as in Figure 4.5, then the total reactive power ripple has the

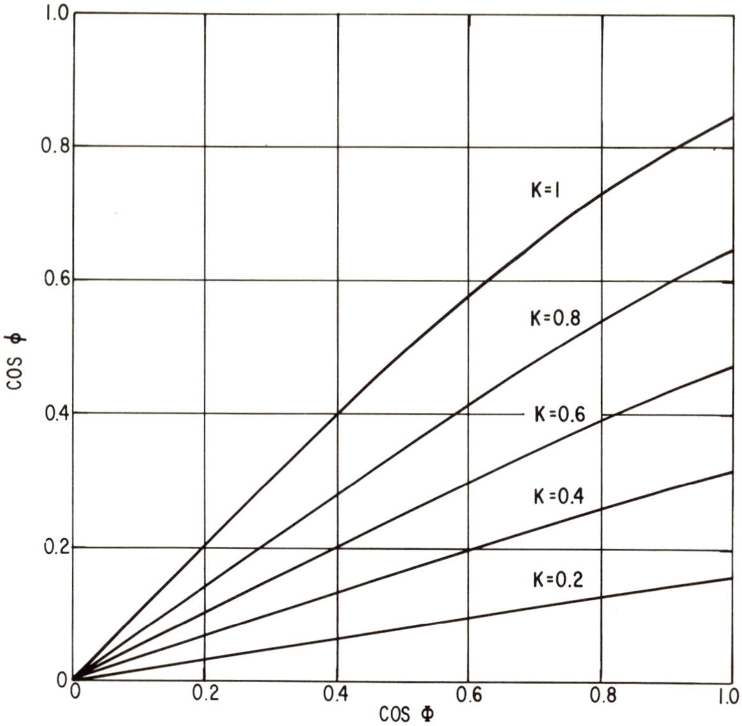

Figure 4.6. Variation of input power factor cos ϕ with output power factor cos Φ and modulation fraction k. Cycloconverter with sinusoidal firing waves and sinusoidal control signal.

same shape for any value of Φ, fluctuating at a frequency $2F\,n_0$ with a min/max ratio of $\sqrt{3}/2$ for $n_0 = 3$. This appears to be the worst-case pulsating load imposed upon the supply and would be approached in practice when starting a motor load at reduced frequency and voltage.

4.4 EQUIVALENT CIRCUIT OF A CYCLOCONVERTER

The same factors that cause loss of voltage in a rectifier circuit are effective in a cycloconverter, and the dc equivalent circuit in Figure 2.2 described by Equation 2.7 is applicable with slight modification. The first change incorporates the contra-phase shift of the opposite polarity groups and provides that the constant drop term E_b always opposes the flow of current. The second change allows that time-varying output current is opposed by inductance in the circuit.

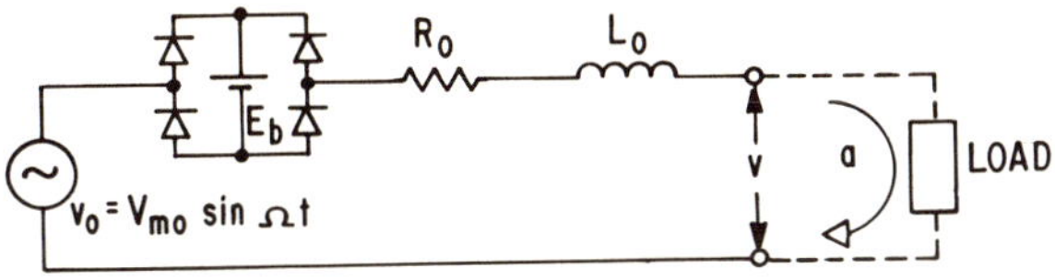

Figure 4.7. Equivalent circuit of a cycloconverter.

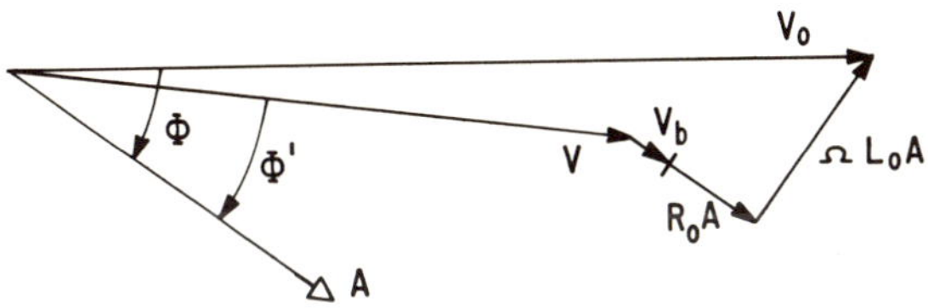

Figure 4.8. Phasor diagram of a cycloconverter.

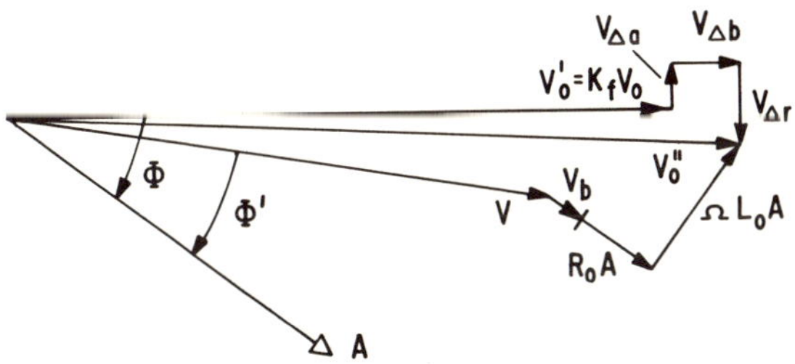

Figure 4.9. Phasor diagram of a cycloconverter, including effect of high frequency ratio. (Note: For the current phase shown, $V_{\Delta b}$ is negative.

Following any possible closed path from source to load and back, it can be seen that the total effective loop inductance L_0 at any instant is

$$L_0 = s(L_c + L_g) + L_f.$$

The contribution of the commutating inductance L_c can also be argued by the method of Amato.[43] The same inductance L_0 should also be added to the rectifier equivalent circuit when it is desired to study transient dc performance.

Incorporating these factors, the cycloconverter equivalent circuit valid for low-frequency modulation ($F \ll f$) may be drawn as in Figure 4.7 and described by the pseudo-instantaneous equation (which is also valid for reversing dc converters),

$$v = V_{m0} \sin \Omega t - E_b \operatorname{sgn}(a) - R_0 a - L_0 \frac{da}{dt}. \tag{4.24}$$

For sinusoidal voltage and current, Equation 4.24 becomes, in rms-phasor notation,

$$V = V_0 - V_b \angle\Phi - R_0 A \angle\Phi - \Omega L_0 A \angle\Phi - 90^\circ,$$

which describes the phasor diagram in Figure 4.8. The small, theoretically square wave voltage E_b is approximated by its fundamental component V_b having the rms value $0.9E_b$. Note that the no-load output voltage or "output emf" V_0 is not measurable in the power circuit but appears as a proportional reference signal in the control circuit when conventional control methods are used. The assumed phase lag Φ of the load current is referenced to this control signal phasor.

However, the load lag is generally specified as the angle Φ' between the terminal voltage V and the current A. Also, it is shown in Chapter 5 that the effect of a high frequency ratio F/f can be approximated by reducing the voltage V_0 by a factor K_f and introducing current-dependent voltage phasors $V_{\Delta a}$, $V_{\Delta b}$, and $V_{\Delta r}$. The magnitude of these voltages can be calculated from Equations 5.22, 5.23, and 5.30, assuming $\Phi \approx \Phi'$. Thus, the phasor diagram becomes as shown in Figure 4.9, where V_0'' is the output emf corrected for the effects of frequency ratio. The magnitude of the terminal voltage V can be calculated as follows:

Let

$$V_0'' = \sqrt{(K_f V_0 + V_{\Delta b})^2 + (V_{\Delta r} - V_{\Delta a})^2} \tag{4.25}$$

$$X_0 = \Omega L_0 \tag{4.26}$$

$$V_J = X_0 A \cos \Phi' - (R_0 A + V_b) \sin \Phi' \tag{4.27}$$

and

$$V_D = X_0 A \sin \Phi' + (R_0 A + V_b) \cos \Phi'. \tag{4.28}$$

Then

$$V = \sqrt{V_0''^2 - V_J^2} - V_D, \tag{4.29}$$

and

$$\Phi = \Phi' + \operatorname{Tan}^{-1} \frac{V_J}{\sqrt{V_0''^2 - V_J^2}} + \operatorname{Tan}^{-1} \frac{V_{\Delta r} - V_{\Delta a}}{K_f V_0 + V_{\Delta b}}. \tag{4.30}$$

To improve the accuracy of the result, these calculations can be iterated using the value of Φ obtained from Equation 4.30 to recalculate the frequency-ratio correction voltages $V_{\Delta a}$ and $V_{\Delta b}$.

4.5 VOLTAGE REGULATION AND THYRISTOR di/dt RATING

In order to derive simple relationships between the circuit and thyristor device parameters, the output voltage expression, Equation 4.24, will be simplified by neglecting the small constant drop E_b and assuming that the output frequency Ω is so low that the equivalent reactance drop can also be neglected. The equivalent resistance drop remains and may be expressed in terms of a regulation factor ϵ defined as the per unit voltage drop with full modulation and unity power-factor load having a specified peak current A_{m1}

$$\epsilon = \frac{R_0 A_{m1}}{E_{do}} = \frac{spf\, L_c A_{m1}}{E_m s(p/\pi) \sin(\pi/p)} = \frac{\pi f\, L_c A_{m1}}{E_m \sin(\pi/p)} \tag{4.31}$$

where the true resistance component of R_0 in equation 2.9 has been neglected. The output power W_0 of a cycloconverter with unity power factor load can be written as

$$W_0 = \frac{n_0(V_{mo} - R_0 A_{m1}) A_{m1}}{2}$$

$$= \frac{n_0 E_{do}^2 (k - \epsilon)\,\epsilon}{2spfL_c}. \tag{4.32}$$

On the basis of system considerations, the circuit designer can set a tolerable upper limit on the value of ϵ. Unfortunately, for applications such as vehicle drives, a limit is difficult to specify without reference to particular cases. The generator voltage E_m and frequency f may both be variable. At maximum motor speed, the cycloconverter output frequency and voltage will be maximum. The neglected reactance drop may now be appreciable and the voltage loss and distortion resulting from high output/input frequency ratio become significant. Under these conditions, an allocation of no more than 5% to the equivalent resistance drop may be desirable.

However, the maximum output current, which specifies the value of J_m for thyristor sizing purposes, occurs when starting the motor, where the output frequency and voltage are minimum. If the supply voltage is maintained relatively high, a regulation factor of up to 50% (defined for $A_{m1} = J_m$) may be tolerable. It will therefore be assumed hereafter that the values of A_{m1} and ϵ pertain to the top-speed, full-load condition and that a constant factor m determined by the drive system requirements relates the two current levels of design significance, that is

$$J_m = mA_{m1}. \tag{4.33}$$

A simple relationship between the di/dt required of the thyristors and the regulation factor is obtained from Equations 2.27 and 4.31, if one assumes that ϵ is defined for the highest supply voltage,

$$\left(\frac{di}{dt}\right)_{\max} = \frac{\pi f A_{m1}}{\epsilon}.$$

For example, if a cycloconverter supplied by a 2 kHz generator must have no more than 5% regulation when delivering 200 amps peak load current, the thyristors must have a di/dt capability of at least 25 amp/μsec.

Any attempt to reduce the regulation by employing thyristors having high di/dt capability is likely to be limited by surge current considerations. The ratio of peak surge current to peak normal current in the thyristors can be found from Equations 4.33, 4.31, 3.4 and 2.25 (for p odd),

$$\frac{J_{sr}}{J_m} = \frac{K_{c2} \cos(\pi/2p)}{2m\epsilon \sin(\pi/p)} \cdot \frac{L_c}{L_c + L_g + M_g}. \tag{4.34}$$

For a typical example, suppose $p = 3$, $K_{c2} = 1.6$, no intergroup reactors are used, and the starting current is $m = 2.5$ times the current at which a regulation factor $\epsilon = 0.05$ is desired. Then,

$$\frac{J_{sr}}{J_m} = \frac{K_{c2}}{2m\epsilon} = \frac{1.6}{2 \times 2.5 \times 0.05} = 6.4.$$

Under the same conditions as the di/dt example, $J_m = 500$ amp and $J_{sr} = 3200$ amp. If the available thyristors do not meet this surge requirement, then higher regulation must be tolerated or intergroup reactors added.

The type of surge rating implied here is based on an offset cosine waveform such as shown in Figure 3.1, with a duration approaching one cycle of the supply frequency (not necessarily 60 Hz, where most present thyristor surge ratings are specified), and presumes that the thyristor regains gate control and will block forward voltage applied after the surge (a relatively long turn-off time is permitted). The surge will be repetitive at some random rate according to the probability of intergroup faults but will not always be initiated at the angle resulting in the worst-case surge.

If the commutating inductance were reduced to the minimum value necessary for limiting di/dt, the surge current would be, for $p = 3$ and no intergroup reactors,

$$J_{sr} = \frac{K_{c2}(di/dt)_{\max}}{\omega}. \tag{4.35}$$

For fast thyristors having high *di/dt* capability, the surge given by Equation 4.35 is generally excessive. However, the snubber discharge current may absorb any *di/dt* capacity not needed for switching load current, as discussed in Section 8.3.

5 Correction Factors for High Frequency Ratio

As the ratio of output to input frequency of a cycloconverter increases, the quality of the output waveform deteriorates. Assuming an open-loop control system, a method of estimating the effective reduction and distortion of the pseudo-instantaneous output voltage will be developed, following the results of Amato.[29,33] The resulting frequency-ratio correction factors can be used to improve the calculation of the net output voltage, as described in Section 4.4. With a closed-loop control system, the voltage reduction can be compensated, within the bandwidth and saturation limits of the converter. Special compensation techniques are required to reduce the distortion.[29] The variation of the pseudo-instantaneous voltage is difficult to discern in the true instantaneous output voltage waveform without filtering, because it is swamped by the ripple at the qth harmonic of the supply frequency.

The analysis begins with an investigation of the angular interval between the firing of one thyristor and the next, assuming a particular control method that is commonly employed. The consequent deviation of the generated output voltage is studied in Section 5.2, where it is shown to be a function of the power factor of the load. Similar distortion caused by variation of the equivalent output resistance is analyzed in Section 5.3. To justify an approximation used when the modulation factor is less than unity, it is shown in Section 5.4 that the results would be exact if a somewhat different method of control were used.

It is often asked, what is the maximum frequency that can be obtained from a cycloconverter? In motor drive systems operating from a 60 Hz source, it may be desired to control the motor over a speed range that requires a frequency approaching or even exceeding 60 Hz. The answer depends upon how poor a waveform can be tolerated by the motor and the susceptibility of the drive to torque pulsations. In Section 5.5, the upper limit of the output frequency relative to the input frequency is discussed. A quantitative measure of the quality of the output waveform is developed in terms of the frequency ratio, converter circuit pulse number, and the modulation fraction. Limitations on the validity of the frequency ratio correction factors are noted.

5.1 VARIATION OF ANGLE BETWEEN SUCCESSIVE FIRING OF THYRISTORS

In a rectifier, the angle θ (at the supply frequency ω) between the firing of one thyristor and the next in sequence is a constant, $2\pi/q$. For a cycloconverter, the cyclic modulation causes the thyristor firings to be closer together or farther apart, depending upon the time in the cycle. Thus, the angle θ varies, by an amount depending upon the ratio of output to input frequency. The variation of θ is analyzed first and then used to obtain corresponding variations in the output voltage and impedance.

The method of control using cosine firing waves is assumed, and the time origin is selected at a point of natural commutation, so that the jth firing wave can be expressed as

$$\cos\left(\omega t - \frac{2j\pi}{q}\right), \tag{5.1}$$

where the range of the firing wave for the positive current group is

$$\frac{2j\pi}{q} < \omega t < \frac{2j\pi}{q} + \pi, \tag{5.2}$$

and for the negative current group,

$$\frac{2j\pi}{q} - \pi < \omega t < \frac{2j\pi}{q}. \tag{5.3}$$

The instantaneous output voltage after the jth firing until the $(j + 1)$th firing follows the wave

$$E_m' \cos\left[\omega t - \frac{(2j+1)\pi}{q}\right], \tag{5.4}$$

where $E_m' = E_m$ and $q = p$ for simple single-way circuits. For sinusoidal modulation, the control signal wave can be expressed as a phase-displaced cosine function with respect to the same time origin,

$$k \cos \frac{\Omega}{\omega}(\omega t - \xi). \tag{5.5}$$

Taking, as an example, a 3-phase single-way circuit ($q = 3$) with a frequency ratio of 1/2, these various waveforms are shown in Figure 5.1 for the positive current group and Figure 5.2 for the negative current group. The instants of firing are determined by the intersections of the control signal wave with the cosine firing waves,

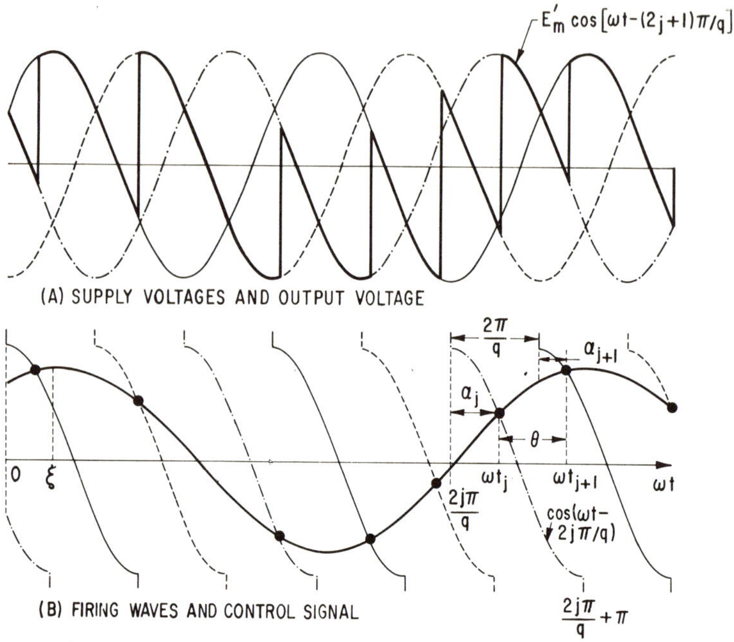

Figure 5.1. Waveforms for positive current group of thyristors, drawn for $q = 3$, $F/f = 1/2$, $k = 0.8$.

shown by dots. If the signal exceeds unity, it will intersect, at either end of the firing waves, the vertical lines that indicate the limit stops built into the control circuit. Thus, the output will be "clipped" if $k > 1$. The time of the jth firing is obtained implicitly by equating Expressions 5.5 and 5.1,

$$k \cos \frac{\Omega}{\omega} (\omega t_j - \xi) = \cos \left(\omega t_j - \frac{2j\pi}{q} \right). \tag{5.6}$$

For the positive current group of thyristors,

$$\omega t_j = \frac{2j\pi}{q} + \alpha_j, \tag{5.7}$$

and

$$\omega t_{j+1} = \frac{2(j+1)\pi}{q} + \alpha_{j+1}. \tag{5.8}$$

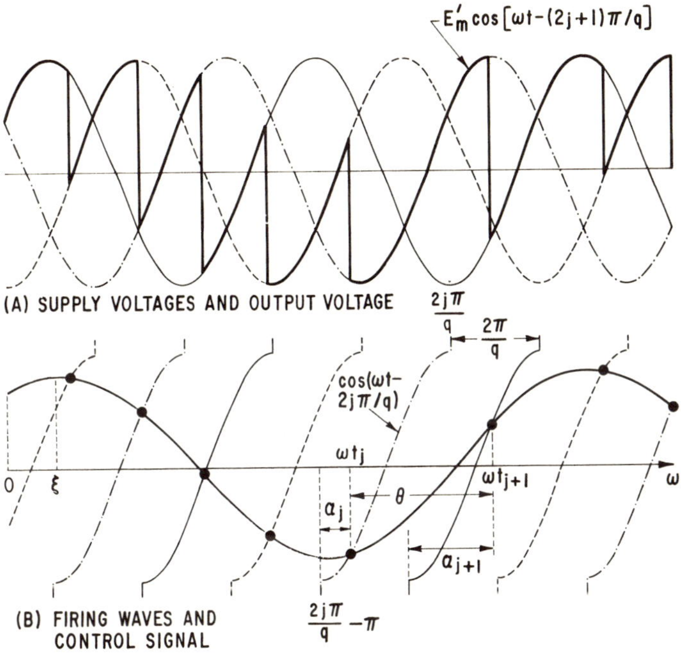

Figure 5.2. Waveforms for negative current group of thyristors, drawn for $q = 3$, $F/f = 1/2$, $k = 0.8$.

The angle θ between successive firings is therefore, by definition,

$$\theta = \omega t_{j+1} - \omega t_j = \frac{2\pi}{q} + \alpha_{j+1} - \alpha_j. \tag{5.9}$$

But, substituting from Equation 5.7 into the right-hand side of Equation 5.6 and solving, one obtains

$$\frac{\Omega}{\omega} (\omega t_j - \xi) = 2m\pi \pm \mathrm{Cos}^{-1} \left[\frac{\cos \alpha_j}{k}\right], \tag{5.10}$$

where m is an integer denoting the number of the output cycle. Here and in the following equations, the top signs apply when the slope of the signal wave is negative or in the first and second quadrants of a cosinusoidal signal (equivalent to the second and third quadrants of a sinusoidal signal). Provided that the slope of the signal remains the same, the next firing occurs with the same value of m and $j = j + 1$ in Equation 5.10. Therefore, taking the difference,

$$\frac{\Omega}{\omega}\,\theta \;=\; \frac{\Omega}{\omega}\,(\omega t_{j+1} - \omega t_j) \;=\; \pm\left[\mathrm{Cos}^{-1}\left(\frac{\cos\alpha_{j+1}}{k}\right) - \mathrm{Cos}^{-1}\left(\frac{\cos\alpha_j}{k}\right)\right]. \tag{5.11}$$

In the special case of unity modulation ($k = 1$), Equation 5.11 simplifies to

$$\frac{\Omega}{\omega}\,\theta \;=\; \pm\,[\alpha_{j+1} - \alpha_j]. \tag{5.12}$$

Then, from Equations 5.9 and 5.12, we obtain

$$\alpha_{j+1} - \alpha_j \;=\; \frac{\Omega/\omega}{\pm 1 - \Omega/\omega}\cdot\frac{2\pi}{q}, \tag{5.13}$$

and

$$\theta \;=\; \frac{1}{1 \mp \Omega/\omega}\cdot\frac{2\pi}{q}. \tag{5.14}$$

For the negative current group of thyristors, the range, Condition 5.3, applies, and Equation 5.7 becomes

$$\omega t_j \;=\; \frac{2j\pi}{q} - \pi + \alpha_j.$$

Following the previous procedure, Equation 5.9 is unchanged, but in Equations 5.10 through 5.14, the $\pm$ signs are reversed. The firing angles of the two polarity groups could have been distinguished by subscripts p or n, as in Chapter 4. However, this would complicate the notation in the present section where subscript j is used.

5.2 VARIATION OF NO-LOAD OUTPUT VOLTAGE

The average output voltage produced by the positive group (neglecting overlap) during the interval between the jth and $(j+1)$th firings is obtained by integrating the instantaneous voltage Expression 5.4 between the limits given by Equations 5.7 and 5.8 and dividing by the interval θ,

$$\begin{aligned} v'_{op} &= \frac{E'_m}{\theta}\int_{2j\pi/q+\alpha_j}^{2(j+1)\pi/q+\alpha_{j+1}} \cos\left[\omega t - \frac{(2j+1)\pi}{q}\right] d(\omega t) \\ &= \frac{E'_m}{\theta}\left[\sin\left(\alpha_{j+1} + \frac{\pi}{q}\right) - \sin\left(\alpha_j - \frac{\pi}{q}\right)\right] \\ &= E'_m\,\frac{2}{\theta}\sin\frac{\theta}{2}\cos\left(\frac{\alpha_j + \alpha_{j+1}}{2}\right). \end{aligned} \tag{5.15}$$

This is a defining equation, valid for any value of k or any type of firing wave or modulating wave.

The desired output voltage during the same interval is the integral of the modulating function, Expression 5.5, between the same limits, multiplied by E_{do}/θ,

$$v_{op} = \frac{E_{do}}{\theta} \int_{\omega t_j}^{\omega t_{j+1}} k \cos \frac{\Omega}{\theta} (\omega t - \xi)\, d(\omega t)$$

$$= \frac{E_{do} k \omega}{\theta \Omega} \left[\sin \frac{\Omega}{\omega} (\omega t_{j+1} - \xi) - \sin \frac{\Omega}{\omega} (\omega t_j - \xi) \right]$$

$$= \pm \frac{E_{do} \omega}{\theta \Omega} \left[\sqrt{k^2 - \cos^2 \alpha_{j+1}} - \sqrt{k^2 - \cos^2 \alpha_j} \right],$$

where the last step is obtained by substitution from Equation 5.10. This result cannot easily be reduced further, except for the special case $k = 1$, for which

$$v_{op1} = \pm \frac{E_{do} \omega}{\theta \Omega} [\sin \alpha_{j+1} - \sin \alpha_j] \tag{5.16}$$

$$= \pm \frac{2 E_{do} \omega}{\theta \Omega} \cos \left(\frac{\alpha_{j+1} + \alpha_j}{2} \right) \sin \left(\frac{\alpha_{j+1} - \alpha_j}{2} \right). \tag{5.17}$$

Dividing Equation 5.15 by 5.17, and then using 2.15, 5.13 and 5.14, one obtains

$$\frac{v'_{op}}{v_{op1}} = \frac{E'_m \sin \dfrac{\theta}{2}}{\pm E_{do} \dfrac{\omega}{\Omega} \sin \left(\dfrac{\alpha_{j+1} - \alpha_j}{2} \right)} \tag{5.18}$$

$$= \frac{\dfrac{\Omega}{\omega} \sin \left(\dfrac{1}{1 \mp \Omega/\omega} \dfrac{\pi}{q} \right)}{\dfrac{q}{\pi} \sin \dfrac{\pi}{q} \sin \left(\dfrac{\Omega/\omega}{1 \mp \Omega/\omega} \dfrac{\pi}{q} \right)}. \tag{5.19}$$

Equation 5.19 gives the ratio of the actual no-load voltage produced by the positive group to the desired value that is programmed by a control signal having 1 per unit sinusoidal modulation. The behavior with reduced modulation ($k < 1$) can be approximated by substituting

$k\Omega/\omega$ in place of Ω/ω alone. It is shown in Section 5.4 that this substitution gives exact results when the control method employs linear firing waves and a triangular modulating signal.

As mentioned previously, the alternative signs in the preceding equations depend upon the slope of the signal and are reversed when the negative group is operating. Thus, if the converter is operated with both groups of thyristors receiving their normally programmed firing signals at all times (that is, there is no suppression or offset bias to reduce the circulating current), then the average output voltage of the two groups does not depend upon the slope of the signal but is simply reduced by a factor K_f, which is a function of the frequency ratio and modulation fraction. Let

$$r = \frac{k\Omega}{\omega} = \frac{kF}{f}.$$

Then v'_{0p}/v_{0p} is approximately Equation 5.19 with r in place of Ω/ω, and the corresponding equation for the negative current group is

$$\frac{v'_{0n}}{v_{0n}} = \frac{r \sin\left(\dfrac{1}{1 \pm r}\dfrac{\pi}{q}\right)}{\dfrac{q}{\pi}\sin\dfrac{\pi}{q}\,\sin\left(\dfrac{r}{1 \pm r}\dfrac{\pi}{q}\right)}.$$

Then

$$K_f = \frac{v'_0}{v_0} = \frac{1}{2}\left(\frac{v'_{0p}}{v_{0p}} + \frac{v'_{0n}}{v_{0n}}\right)$$

$$= \frac{r}{2\dfrac{q}{\pi}\sin\dfrac{\pi}{q}}\left[\frac{\sin\left(\dfrac{1}{1-r}\dfrac{\pi}{q}\right)}{\sin\left(\dfrac{r}{1-r}\dfrac{\pi}{q}\right)} + \frac{\sin\left(\dfrac{1}{1+r}\dfrac{\pi}{q}\right)}{\sin\left(\dfrac{r}{1+r}\dfrac{\pi}{q}\right)}\right]. \qquad (5.20)$$

For calculations based upon phasor representation of an equivalent circuit, the rms magnitude V_0 of the no-load output voltage can be reduced by the factor K_f,

$$V'_0 = K_f V_0.$$

The difference between the voltages produced by the two polarity groups causes variation of the circulating current in converters having no suppression. This difference may be defined as $2K_\Delta v_0$, where

$$K_\Delta = \frac{1}{2}\left(\frac{v'_{0p}}{v_{0p}} - \frac{v'_{0n}}{v_{0n}}\right)$$

$$= \pm \frac{r}{2\frac{q}{\pi}\sin\frac{\pi}{q}}\left[\frac{\sin\left(\frac{1}{1-r}\frac{\pi}{q}\right)}{\sin\left(\frac{r}{1-r}\frac{\pi}{q}\right)} - \frac{\sin\left(\frac{1}{1+r}\frac{\pi}{q}\right)}{\sin\left(\frac{r}{1+r}\frac{\pi}{q}\right)}\right], \qquad (5.21)$$

where, as before, the alternative signs depend upon the slope of the signal. In converters having suppression of circulating currents, a distortion voltage v_Δ of magnitude $K_\Delta v_0$ adds to or subtracts from the voltage $K_f v_0$ according to the polarity of the load current, which is mathematically described by $\mathrm{sgn}(a)$.

$$v'_0 + v_\Delta = [K_f + K_\Delta \,\mathrm{sgn}\,(a)]\, v_0.$$

In Figure 5.3 (A), the voltages v'_{0p} and v'_{0n} produced by the two polarity groups are shown, together with the ideal output voltage v_0. The components $v'_0 = K_f v_0$ and $K_\Delta v_0$ are drawn in Figure 5.3 (B), from which the distortion component $v_\Delta = K_\Delta \,\mathrm{sgn}\,(a)\, v_0$ can be deduced, as shown in Figure 5.4, for purely resistive, inductive, and capacitive loads. It is seen that the phase of the fundamental component (which may not be regarded as distortion) rotates in the opposite direction as the phase of the current is rotated. For purely reactive load, there are no harmonic distortion terms. The situation for a current phase lag Φ is depicted in Figure 5.5 (B), and the fundamental distortion term can be resolved into rms components V_{Δ_b} and V_{Δ_a}, which are, respectively, in phase with the signal voltage V_0 and leading it by 90°. These components are given by the following Fourier analysis:

$$V_{\Delta a} = \frac{2}{\sqrt{2}\pi}\int_{-\frac{\pi}{2}}^{\frac{\pi}{2}} [K_\Delta \,\mathrm{sgn}\,(a)\, V_{m0} \sin x] \cos x \, dx$$

$$= \frac{2}{\pi} K_\Delta V_0 \left[\int_{-\frac{\pi}{2}}^{\Phi} (-\sin x) \cos x \, dx + \int_{\Phi}^{\frac{\pi}{2}} (\sin x) \cos x \, dx\right]$$

$$= \frac{2}{\pi} K_\Delta V_0 \cos^2 \Phi. \qquad (5.22)$$

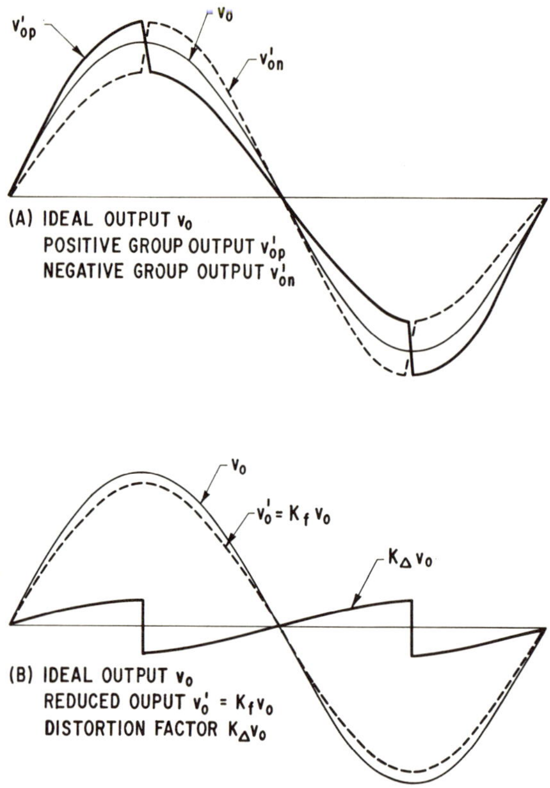

Figure 5.3. Effect of high frequency ratio on output voltage of cycloconverter.

For $V_{\Delta b}$, the multiplying factor in the integrand becomes sin x instead of cos x, and the result is

$$V_{\Delta b} = \frac{1}{\pi} K_{\Delta} V_0 \ (\sin 2\Phi - 2\Phi). \tag{5.23}$$

Equations 5.22 and 5.23 are valid for $-\frac{\pi}{2} < \Phi < \frac{\pi}{2}$. When the current is reversed, as shown in Figure 5.5 (C), the phases of $V_{\Delta a}$ and $V_{\Delta b}$ are reversed also.

Values of the voltage reduction factor K_f and the voltage distortion factor K_{Δ} are plotted in Figure 5.6 as a function of $r = k\ F/f$ for pulse numbers $q = 2$, 3, 4, 6, and 12. These frequency ratio correction factors become meaningless when $K_f = |K_{\Delta}|$, which implies that the voltage v_0' is reduced to zero for part of a cycle. Examination of

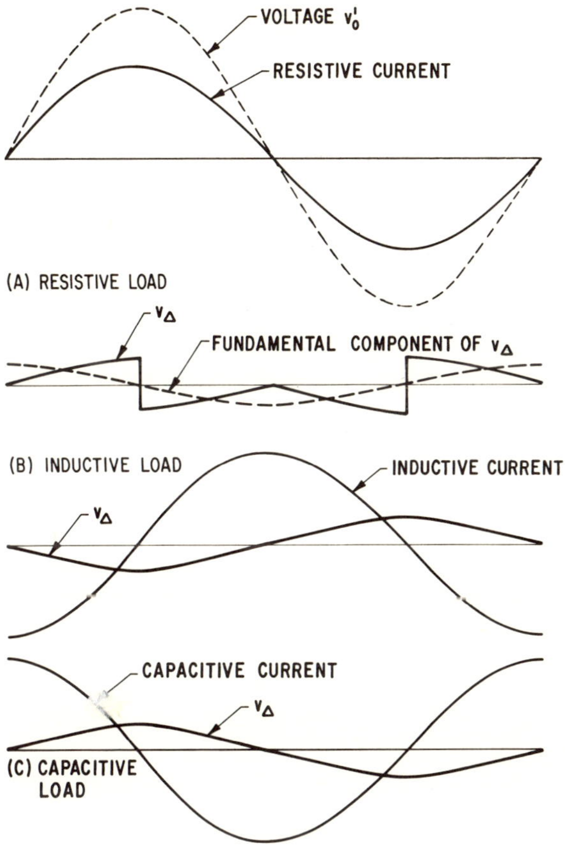

Figure 5.4. Distortion voltage v_Δ caused by high frequency ratio for different loads.

Equation 5.18 reveals that this condition is reached when $\sin\theta/2 = 0$, or

$$\frac{\theta}{2} = \frac{1}{1-r}\frac{\pi}{q} = \pi, \tag{5.24}$$

which results in the following restriction of the ratio factor r,

$$r \leqslant \left(\frac{q-1}{q}\right). \tag{5.25}$$

However, the limiting value of θ given by Equation 5.24 can only be attained for $q = 2$, because the firing angle α is generally restricted

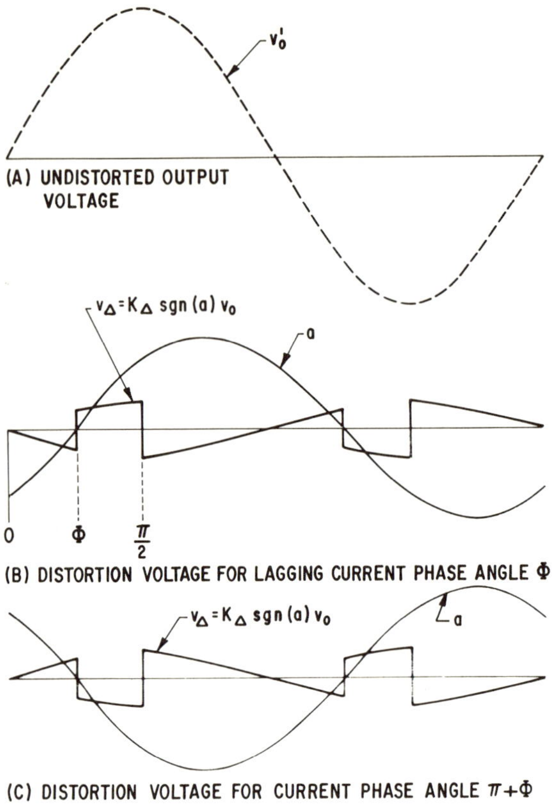

Figure 5.5. Distortion voltage $v\Delta$ for general type of load.

to the range $0 < \alpha < \pi$. If one assumes that one thyristor fires at the advance limit $\alpha = 0$ and the next at the retard limit $\alpha = \pi$, the maximum possible interval between successive thyristor firings is

$$\theta = \pi + \frac{2\pi}{q},$$

and the corresponding restriction on the ratio r is

$$r \leqslant \left(\frac{q}{q+2}\right). \tag{5.26}$$

The loci of these restrictions are indicated in Figure 5.6. The modulating signal does not necessarily have to observe these limitations, but the factors K_f and $K\Delta$ do not accurately describe the resulting

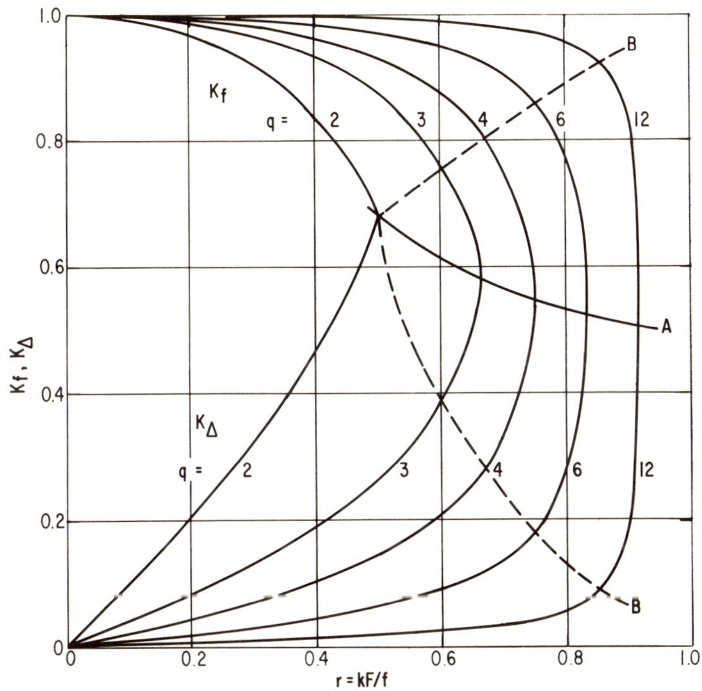

Figure 5.6. Voltage reduction factor K_f and distortion factor K_Δ as a function of frequency ratio factor r and converter pulse number q. The loci of frequency ratio factor restrictions are indicated:
(A) $r \leqslant (q-1)/q$;
(B) $r \leqslant q/(q+2)$.

output when Condition 5.26 is violated and become numerically ridiculous when Condition 5.25 is violated.

5.3 VARIATION OF EQUIVALENT RESISTANCE

The equivalent output resistance R_0, Equation 2.9, is derived from the voltage reduction due to commutation overlap, Equation 2.4. Written with the pseudo-instantaneous variables pertaining to a cycloconverter, Equation 2.4 becomes

$$v_x = s\,pf\,[L_c a + L_u I_s \operatorname{sgn}(a)]. \tag{5.27}$$

The term in brackets represents the flux linkages (volt seconds) lost per commutation, which is multiplied by the number of commutations per unit time ($s\,pf$). Now, the latter is inversely proportional to the angle θ between commutations, so that as θ varies about its dc value $2\pi/q$, Equation 5.27 must be multiplied by the factor

$$\frac{2\pi/q}{\theta} = 1 \mp \frac{k\Omega}{\omega}, \tag{5.28}$$

where θ has been obtained from Equation 5.14 with the approximate allowance for less than unity modulation. The second term in Equation 5.28 generates a distortion voltage $v_{\Delta r}$, and the dependence of the alternative signs upon the voltage slope and the current polarity can be expressed mathematically using the sign function,

$$\begin{aligned} v_{\Delta r} &= \operatorname{sgn}\left(\frac{dv}{dt}\right) \operatorname{sgn}(a) \frac{k\Omega}{\omega} s\, pf\,[L_c a + L_u I_s \operatorname{sgn}(a)] \\ &= \operatorname{sgn}\left(\frac{dv}{dt}\right) k\, s\, pF\,[L_c |a| + L_u I_s]. \end{aligned} \tag{5.29}$$

The behavior of the voltage $v_{\Delta r}$ for different types of load is illustrated in Figure 5.7. Note that the fundamental component of $v_{\Delta r}$ leads the signal voltage v_0 for any current phase angle. Therefore, the effect of the equivalent resistance variation can be represented in phasor calculations by adding a phasor $V_{\Delta r}$, which lags the phasor V_0 by 90°. The magnitude of $V_{\Delta r}$ is proportional to the peak current A_m for purely reactive load and $2A_m/\pi$ for purely resistive load. As an approximation, an average factor of 0.8 can be used in calculating the rms voltage $V_{\Delta r}$,

$$V_{\Delta r} = k\, s\, pF(0.8 L_c A + 0.9 L_u I_s). \tag{5.30}$$

5.4 FREQUENCY RATIO EFFECT WITH LINEAR FIRING WAVES AND LINEAR MODULATION

This particular variation of the control method leads to a simple accounting of the frequency ratio correction terms when the modulation factor k is less than unity. When $k = 1$, the method produces the same sinusoidal output voltage as the conventional method that employs cosine firing waves and sinusoidal modulation. As k becomes smaller, the output deteriorates into a triangular waveform. However, the use of the results of the following analysis as an approximation in the case of the conventional (or other) control methods is justified because the correction terms have a relatively small effect over most of the frequency range where they are valid.

To simplify the discussion, only the positive current group will be considered. The jth firing wave may be expressed as

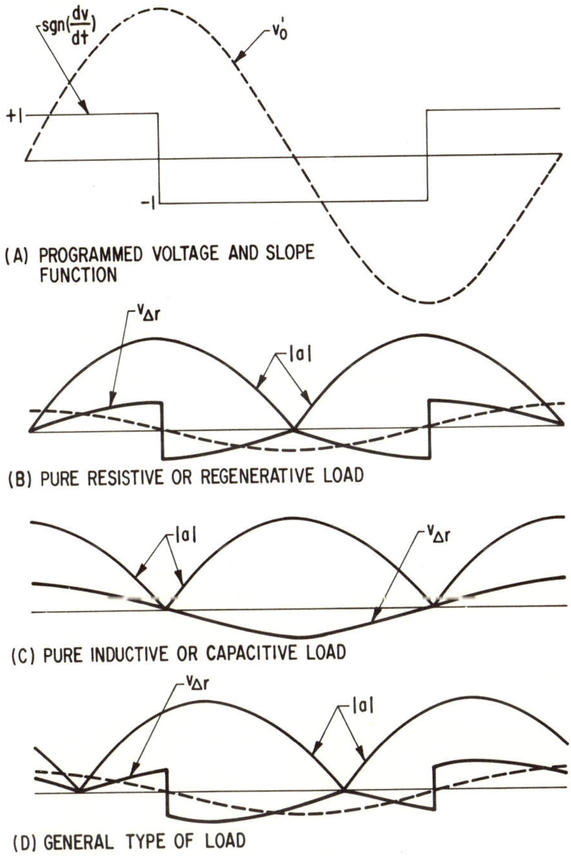

Figure 5.7. Distortion voltage caused by variation of equivalent resistance. In (B) and (D) the fundamental component is shown by the dashed line.

$$-\frac{2}{\pi}\left(\omega t - \frac{2j\pi}{q}\right) \tag{5.31}$$

over the range

$$\frac{2j\pi}{q} - \frac{\pi}{2} < \omega t < \frac{2j\pi}{q} + \frac{\pi}{2}.$$

With appropriate selection of its phase with respect to the time origin, the signal wave can be written as

$$(-1)^m \, k\frac{2}{\pi}\,(\Omega t - m\pi) \tag{5.32}$$

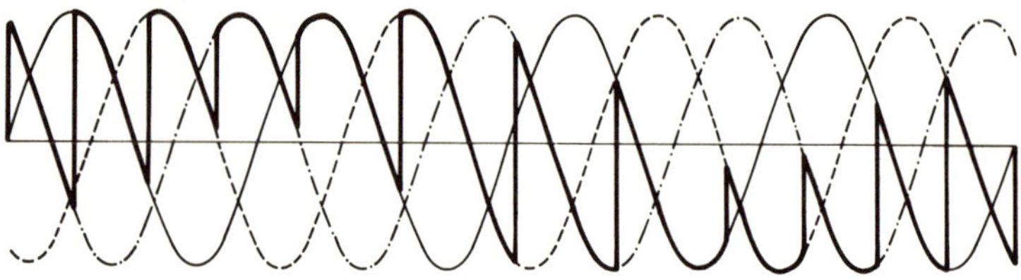

(A) SUPPLY VOLTAGES AND OUTPUT VOLTAGE

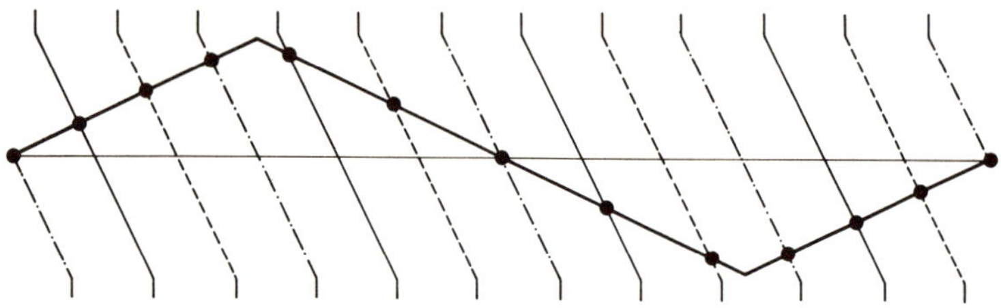

(B) FIRING WAVES AND CONTROL SIGNAL

Figure 5.8. Waveforms with linear firing waves and linear modulation (triangular control signal). Drawn for positive current group of thyristors, $q = 3$, $F/f = 1/4$, $k = 1$.

over the range

$$m\pi - \frac{\pi}{2} < \Omega t < m\pi + \frac{\pi}{2},$$

where m is an integer having odd values when the slope is negative and even values when the slope is positive. These waveforms are sketched in Figure 5.8 for the case of a 3-pulse circuit with a frequency ratio of 1/4. The firing points are the intersections of the signals given by Expressions 5.31 and 5.32, and the jth firing at time t_j occurs when

$$(-1)^m k \frac{2}{\pi} (\Omega t_j - m\pi) = -\frac{2}{\pi}\left(\omega t_j - \frac{2j\pi}{q}\right)$$

$$= 1 - \frac{2}{\pi} \alpha_j \qquad (5.33)$$

because

$$\omega t_j = \frac{2j\pi}{q} - \frac{\pi}{2} + \alpha_j.$$

If the value of m remains the same, the $(j+1)$th firing is given by Equation 5.33 with $j = j + 1$. The difference in time between the two firings is then given by

$$(-1)^m \ k \ \frac{2}{\pi} (\Omega t_{j+1} - \Omega t_j) \ = \ - \ \frac{2}{\pi} (\alpha_{j+1} - \alpha_j)$$

or

$$k \frac{\Omega}{\omega} \theta \ = \ (-1)^{m+1} (\alpha_{j+1} - \alpha_j),$$

which is the same as Equation 5.12 with Ω/ω replaced by $k \ \Omega/\omega$.

For low output frequency, the firing angle modulation is best described in terms of the angle ψ defined by Equations 4.6 and 4.7,

$$\psi \ = \ (-1)^m \ k \ (\Omega t - m\pi).$$

Thus, the expected pseudo-instantaneous output voltage is

$$v_0 \ = \ E_{do} \sin \psi \ \doteq \ (-1)^m \sin k \ (\Omega t - m\pi),$$

and the average value during the interval between the jth and $(j+1)$th firing is

$$v_0 \ = \ \frac{E_{do}}{\theta} \int_{\omega t_j}^{\omega t_{j+1}} (-1)^m \sin k \ \frac{\Omega}{\omega} \left(\omega t - \frac{\omega}{\Omega} m\pi \right) d(\omega t)$$

$$= \ \frac{E_{do}\omega}{\theta \ k\Omega} (-1)^{m+1} [\cos \psi_{j+1} - \cos \psi_j]$$

$$v_{op} \ = \ (-1)^{m+1} \frac{E_{do}\omega}{\theta \ k\Omega} [\sin \alpha_{j+1} - \sin \alpha_j],$$

which is the same as Equation 5.16 with Ω/ω replaced by $k \ \Omega/\omega$. Similarly, the other results of Section 5.2 obtained for the special case of $k = 1$ are generally applicable to the present type of control when this substitution is made. Conversely, operation of the conventional control method with $k < 1$ can be described approximately by the substitution.

5.5 FREQUENCY RATIO LIMITATIONS

The ratio of the maximum output frequency of a cycloconverter to the supply frequency depends upon the pulse number q of the circuit, the modulation factor k, and the amount of distortion that can be tolerated in a particular application, which is subject to considerable debate. Theoretical criteria will be presented here to roughly define the amount of distortion. Another important factor, not considered quantitatively here, is whether circulating currents are permitted or suppressed. If they are suppressed by blocking the firing pulses of the group of thyristors not carrying load current, then there is a delay between the extinction of current in one group and the first firing of a thyristor in the incoming group. The minimum delay is equal to the turn-off time of the thyristors, and the maximum delay may extend up to the angle θ between successive firings if the firing control logic is not carefully designed. If gate signals are applied to both groups of thyristors at all times, which causes circulating current to flow, then the crossover should be much smoother because a thyristor in the incoming group will be ready to pick up the load current as it reverses.

With sufficiently low modulation, the "ultimate" output frequency can be considered to be that for which every thyristor firing produces alternate polarity output. There are qf pulses per second, and to produce an output of frequency F at least $2F$ pulses per second are required. Therefore,

$$\left(\frac{F}{f}\right)_{\max} = \frac{q}{2}. \tag{5.34}$$

At this limit condition, the output depends upon the phase of the control signal relative to the supply and can be zero. For slightly lower output frequency, there will be severe beat frequency modulation. Hence, operation here is not very practical, unless the frequencies are locked and the circuit is operated as a frequency multiplier.[66]

When three converters are used to provide 3-phase output, a balanced condition at the frequency ratio given by Equation 5.34 can be obtained if the supplies for the three converters are phase displaced by $2\pi/3q$ radians. However, if the same q-pulse supply phases are used for the three converters, then a half-wave balanced condition requires that $F/f = q/3$, or for full-wave balance, $F/f = q/6$. When the acutal frequency ratio is close but not locked to such rational fractions, then severe beat frequency modulation is expected.

A practical upper limit for the modulation fraction k can be established by assuming the criterion that the signal wave should not intersect any firing wave at more than one point. Of course, if more than one intersection does occur, the first will trigger a thyristor and others should be ignored. However, as the waves drift with respect to each other, a condition will occur where the firing point suddenly jumps from one angle to another. Such discontinuous beat frequency distortion is probably undesirable, and therefore we may require that the slope of the control signal wave never exceed the slope of the firing waves. If the shapes of these waves are the same (for example, both cosine or both linear), then the criterion may be stated as

$$\frac{F}{f} < \frac{1}{k}. \tag{5.35}$$

Since $k \leqslant 1$ this restriction is applicable only when an attempt is made to produce a frequency higher than the supply. However, the criterion becomes effective at a lower output frequency when a flat-topped output is programmed.

Further restrictions on the frequency ratio as a function of both q and k can be derived from the theory of the "correction factors," described previously. This theory gives a quantitative basis for specifying the amount of distortion. Condition 5.26 is more stringent than Condition 5.35, but exceeds the limit of Condition 5.34 for low values of k and may be too restrictive for high values of k. An attempt will therefore be made to derive a more comprehensive frequency ratio limit criterion that provides a practical measure of the quality of the output waveform.

The correction factors are calculated on the basis that the angle θ between successive thyristor firings is less for one output half-cycle than for the other. However, the concept of an "angle between successive firings" seems meaningless when there can be no more than one thyristor firing during the half-cycle in which expansion of θ occurs. An output half-cycle lasts for π radians at frequency F, or $\pi f/F$ at frequency f. A criterion for the limit of usable waveshape can be specified by requiring that there be at least one firing during this interval, or that the expanded value of θ be less than the interval,

$$\theta = \frac{1}{1 - kF/f} \cdot \frac{2\pi}{q} < \pi \frac{f}{F}$$

or

$$\frac{F}{f} < \frac{1}{k + 2/q}.$$

A better criterion, considering that the waveform is improved during the half-cycle in which the thyristor firings are closer together, is to use the geometric mean θ_{gm} of the expanded and contracted values of θ. Extending the argument, a criterion for the "Nth degree of good waveform" is to specify that $N\theta_{gm}$ should be less than an output half-cycle:

$$N\theta_{gm} = \frac{N}{\sqrt{(1 - kF/f)(1 + kF/f)}} \frac{2\pi}{q} < \pi \frac{f}{F}$$

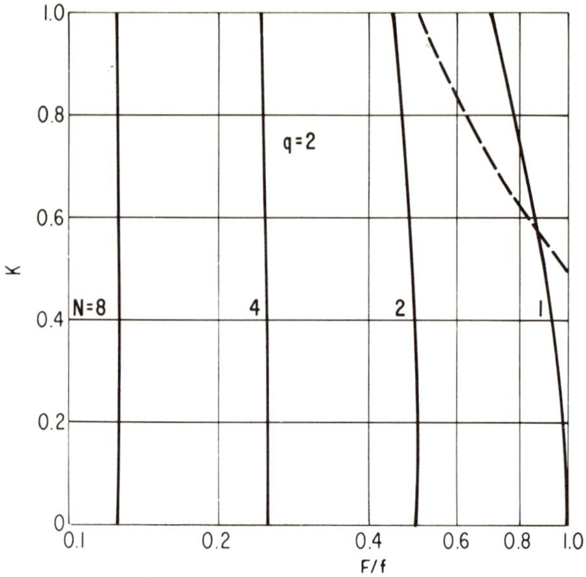

Figure 5.9. Output waveform quality regions as a function of frequency ratio F/f and modulation fraction k. Circuit pulse number $q = 2$.

or
$$\frac{F}{f} < \frac{(q/2N)}{\sqrt{1 + k^2(q/2N)^2}}. \tag{5.36}$$

The right-hand side of Inequality 5.36, for $N = 1$, approaches but does not exceed the limits set by Equation 5.34 and Condition 5.35.

The boundaries established by this criterion with $N = 1, 2, 4,$ and 8 are drawn in Figures 5.9 to 5.13 for $q = 2, 3, 4, 6,$ and 12, respectively; and the regions can be described, somewhat arbitrarily, as good, fair, poor, and bad (from left to right). However, there appears to be some consensus among those who have experimentally evaluated cycloconverters that these distinctions are reasonable. The relative importance of k and q in determining the output distortion is well illustrated. Converters for high-quality power supplies, such as aircraft VSCF (variable speed, constant frequency) systems, should operate in the "good" region. Motor drives can operate quite efficiently into the "fair" region. Operation in the "poor" region should not be considered unless other means of providing the application requirements are equally unattractive. Inequality 5.36 can be rearranged with $N = 2$ to place a restriction on the modulation

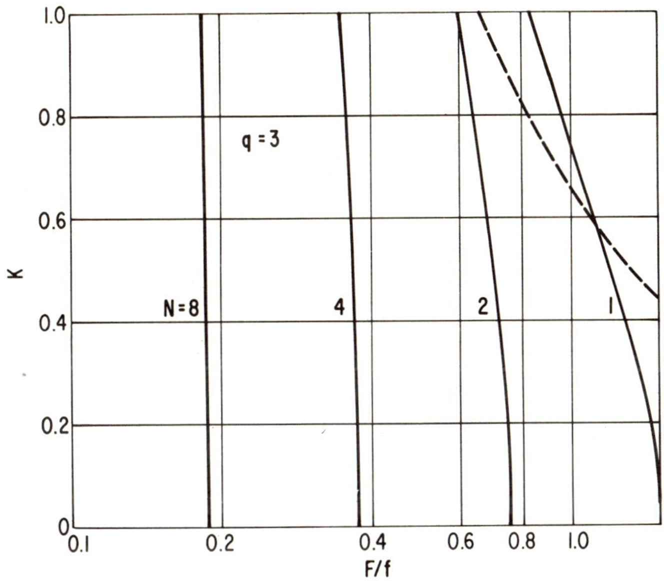

Figure 5.10. Output waveform quality regions as a function of frequency ratio F/f and modulation fraction k. Circuit pulse number $q = 3$.

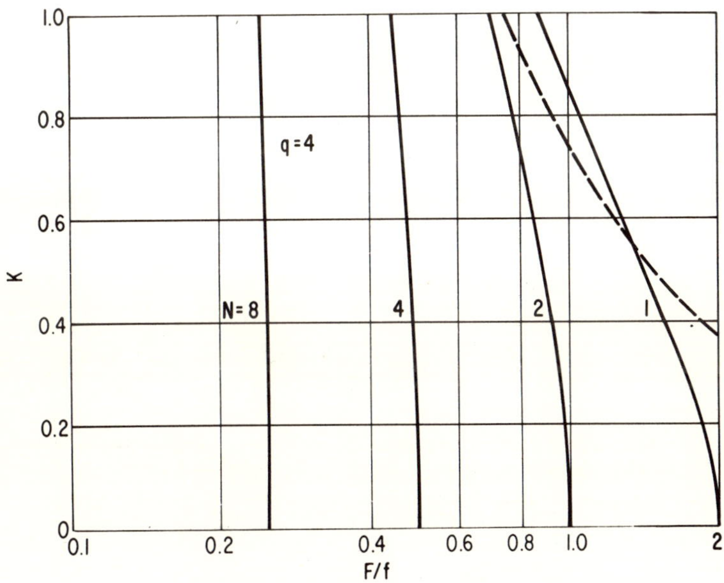

Figure 5.11. Output waveform quality regions as a function of frequency ratio F/f and modulation fraction k. Circuit pulse number $q = 4$.

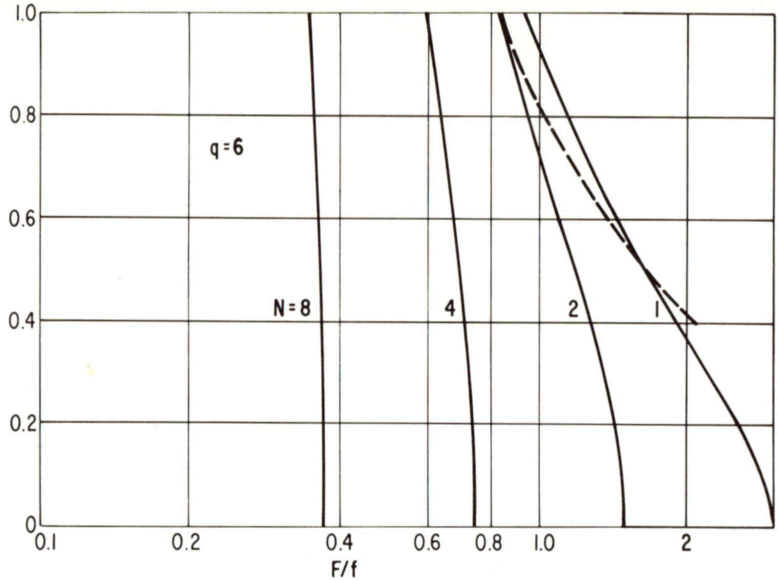

Figure 5.12. Output waveform quality regions as a function of frequency ratio F/f and modulation fraction k. Circuit pulse number $q = 6$.

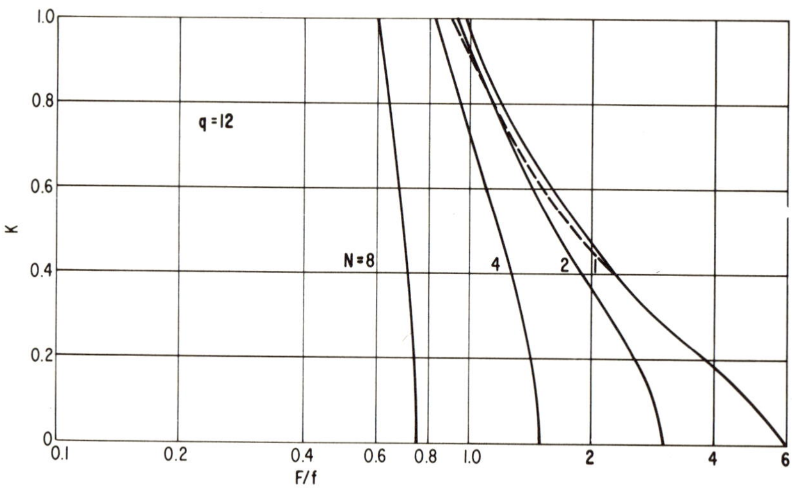

Figure 5.13. Output waveform quality regions as a function of frequency ratio F/f and modulation fraction k. Circuit pulse number $q = 12$.

factor k when the pulse number and desired frequency ratio are given,

$$k < \sqrt{\left(\frac{f}{F}\right)^2 - \left(\frac{4}{q}\right)^2}. \tag{5.37}$$

If the correction factors K_f and K_Δ given by Equations 5.20 and 5.21, respectively, are used to compute the output voltage, then Restriction 5.25 must not be violated. This limit is indicated by dashed lines in Figures 5.9 to 5.13 and is seen to conflict with Condition 5.37 only when $q > 6$. As the frequency ratio is raised toward these limits, the random nature of cycloconverter operation becomes more noticeable (each output cycle is unlike any other unless the frequency ratio happens to be a rational number), and the validity of the continuous-function correction factors becomes questionable. However, they appear to be reasonably accurate for "good" or "fair" waveforms, $N \geqslant 4$ in Figures 5.9 to 5.13 and may be used up to the limit set by Restriction 5.25 as a practical approximation.

For a given waveform quality with full modulation, the value of the voltage reduction factor K_f is approximately constant, independent of the pulse number. If the value of the frequency ratio F/f for $k = 1$ at the boundary between the "good" and "fair" regions ($N = 8$) in Figures 5.9 to 5.13 is entered into Figure 5.6, the value of K_f is found to be about 0.99 for all values of q. Similarly, at the boundary between the "fair" and "poor" regions ($N = 4$), it is noted that K_f is reduced to about 0.95. The distortion factor K_Δ is, however, more dependent on the pulse number. For the common 6-pulse converter, the corresponding values are $K_\Delta \approx 0.04$, for $N = 8$, and $K_\Delta \approx 0.09$, for $N = 4$.

6 Nonsinusoidal Operation

Usually, ac power apparatus is designed to perform most efficiently with a sinusoidal supply voltage waveform. Some equipment requires a sine wave for proper operation or must be derated if large harmonic components are present. For this reason, cycloconverters used as general purpose power supplies are required to meet low-harmonic waveform specifications. However, in special purpose applications, it may be advantageous to compromise the ideal waveform to improve the performance of the system in other respects.

For example, in large motor-drive systems, the input power factor is very important, because it affects the utility charges. The power factor of a dc converter is highest when the firing angle is fully advanced (rectifier mode) or fully retarded (inverter mode). The power factor of a sine-wave cycloconverter is highest when the modulation factor is unity. Both the power factor and output voltage can be increased by overmodulation or by programming a flat-topped waveform such as a trapezoid. This causes the firing angle to be fully advanced or fully retarded for a greater portion of each cycle, instead of only at the peaks. The resulting harmonics can sometimes be reduced by suitable output connections.

Nonsinusoidal output may also be adopted primarily as a practical expedient, because it can be produced with simpler, less costly control circuitry. This factor is of relatively greater importance in small equipment, such as drives for small motors. When it also yields better thyristor utilization and higher power factor, the attractiveness of the scheme is enhanced.

The factors that can contribute to the generation of nonsinusoidal output are surveyed in Section 6.1 and analyzed in Section 6.2. Calculation of the input power factor when the output is not a sine wave is described in Section 6.3. Extension of this calculation to include the effect of commutation overlap, whether the output is sinusoidal or not, is discussed in Section 6.4.

A technique for improving the power factor of a cycloconverter without necessarily distorting the programmed output waveform is studied in Section 6.5. The method is applicable to converters having at least two commutating groups and involves unsymmetrical control of the groups. The portion of the output voltage generated by each group is severely distorted, but the combined output can be sinusoidal or may still be flat-topped to further increase the power factor. This unsymmetrical firing technique has been used to improve the power factor of converters with dc output and is an extension of the half-thyristor, half-diode "semiconverter" rectifier circuit.

6.1 FACTORS AFFECTING PROGRAMMED VOLTAGE WAVEFORM

Generation of a pseudo-sinusoidal output voltage is described in the preceding sections when the supply voltage, firing wave, and control signal are all sinusoidal. Operation when some of these waveforms are not sinusoidal will now be considered.

6.1.1 Supply Voltage

Other than undesired system distortion, a practical case of a nonsinusoidal supply would be a square wave produced by an inverter. The cascade connection of an inverter and a cycloconverter has been named a "cycloinverter" by some authors. Two stages of power conversion are usually less efficient than direct inversion to the final output frequency, using some other method to shape the waveform. However, a high frequency inverter allows a small transformer to be used for voltage matching, and an inverter-cycloconverter system has been proposed where the size of a transformer at the output frequency would be prohibitive.

6.1.2 Firing Wave

In the usual method of cycloconverter control, the firing angle is determined by comparing the control signal with a "firing wave." A thyristor trigger pulse is generated when the two signals intersect. If the firing wave voltage is proportional to the cosine of the prospective firing angle α (or the sine of the angle ψ), then the pseudo-instantaneous output voltage is directly proportional to the instantaneous control signal voltage, and the converter behaves as a linear amplifier. This is a result of Equation 2.2 or, more fundamentally, Equation 5.15 where it is assumed that the supply is sinusoidal. In general, the fundamental requirement for a linear transfer function is that the firing wave be proportional to the integral of the supply voltage. If the supply is a square wave, then a linear firing wave (voltage $\propto$ ψ) produces a linear transfer function. Conversely, using a linear firing wave with a sinusoidal supply results in a sinusoidal transfer function (output voltage $\propto$ sine of signal). Since a linear firing wave may be practically convenient or desirable, this case will be studied in Section 6.2.

6.1.3 Control Signal

While most applications desire sinusoidal output, it is generally easier to produce certain types of nonsinusoidal control signal reference waveforms, particularly when polyphase output is required over a wide frequency range. A flat-topped waveform has a higher ratio of fundamental component to peak voltage, resulting in better utilization of the thyristors and improved input power factor. In a 3-phase system where three converters can be connected in wye, the triple harmonics in such a flat-topped wave are neutralized and do not appear across the the load.

6.2 GENERAL ANALYSIS

The following definitions in per-unit terms are used here:
Control signal voltage

$$v_c = f(\Omega t)$$

Sinusoidal firing wave voltage

$$v_{f1} = \sin \psi$$

Linear firing wave voltage

$$v_{f2} = \frac{2}{\pi} \psi$$

Output voltage

$$\bar{v}_0 = v_0 / E_{do} = \sin \psi \tag{6.1}$$

Thus, for sinusoidal firing waves,

$$\psi = \mathrm{Sin}^{-1}(v_c) = \mathrm{Sin}^{-1} f(\Omega t)$$

$$\bar{v}_0 = v_c \tag{6.2}$$

and, for linear firing waves,

$$\psi = \frac{\pi}{2} v_c = \frac{\pi}{2} f(\Omega t)$$

$$\bar{v}_0 = \sin\left(\frac{\pi}{2} v_c\right) = \sin\left[\frac{\pi}{2} f(\Omega t)\right]. \tag{6.3}$$

Examples given by Bland[57] include sinusoidal, triangular, and square-wave control signals combined with sinusoidal or linear firing

Table 6.1. Summary of Control Methods.

Control Method No.	Control Waveforms Used with Sinusoidal Supply: Control signal	Firing wave $-\frac{\pi}{2} < \psi < \frac{\pi}{2}$	Per-Unit Equations: Control signal $v_c = f(\Omega t)$	Per-Unit Equations: Output Voltage $\bar{v}_0$	Output Voltage Parameters: Fundamental Magnitude a_1	a_1 for $k = 1$	Per-Unit nth (ODD) Harmonic $\frac{a_n}{a_1}$	Total rms Harmonics (1-Phase) $\sqrt{\sum\left(\frac{a_n}{a_1}\right)^2}$ $n = 3,5,7,9$	Total rms Harmonics (3-Phase) $\sqrt{\sum\left(\frac{a_n}{a_1}\right)^2}$ $n = 5,7,11,13$	Maximum Average Displacement Power Factor of Input (Obtained when $k = 1$ and $\cos\Phi = 1$)
1	Sinusoidal	$\sin\psi$	$k \sin \Omega t$		k	1	0	0	0	0.844
2	Triangular	$\sin\psi$	$(-1)^m k \frac{2}{\pi}(\Omega t - m\pi)$ $m\pi - \pi/2 < \Omega t < m\pi + \pi/2$		$\frac{8}{\pi^2} k$	0.811	$\frac{1}{n^2}$	0.121	0.046	0.675
3	Square	$\sin\psi$	$(-1)^m k,\ m\pi < \Omega t < (m+1)\pi$		$\frac{4}{\pi} k$	1.273	$\frac{1}{n}$	0.483	0.310	1
4	Sinusoidal	$\frac{4}{\pi}\psi$	$k \sin \Omega t$	$\sin\left(\frac{k\pi}{2}\sin \Omega t\right)$	$2J_1\left(\frac{k\pi}{2}\right)$	1.134	$\frac{J_n(k\pi/2)}{J_1(k\pi/2)}$	0.122/0*	0.004/0*	0.946
5	Triangular	$\frac{2}{\pi}\psi$	$(-1)^m k \frac{2}{\pi}(\Omega t - m\pi)$	$(-1)^m \sin k(\Omega t - m\pi)$	$\frac{4k\cos(k\pi/2)}{\pi(1-k^2)}$	1	$\frac{1-k^2}{n^2-k^2}$	0/0.121*	0/0.046*	0.844
6	Square	$\frac{2}{\pi}\psi$	$(-1)^m k$	$(-1)^m \sin\frac{k\pi}{2}$	$\frac{4}{\pi}\sin\frac{k\pi}{2}$	1.273	$\frac{1}{n}$	0.483	0.310	1
7	Sinusoidal + third harmonic	$\sin\psi$	$k\frac{2}{\sqrt{3}}\left(\sin \Omega t + \frac{\sin 3\Omega t}{6}\right)$		$\frac{2}{\sqrt{3}} k$	1.155	$\frac{a_3}{a_1} = \frac{1}{6}$	0.167	0	0.947
8	Trapezoidal (c is the shape parameter)	$\sin\psi$	$(-1)^m \frac{2k}{\pi c}(\Omega t - m\pi)$ $0 < c < 1$ $m\pi - c\pi/2 < \Omega t < m\pi + c\pi/2$		$\frac{8k}{\pi^2 c}\sin\frac{c\pi}{2}$		$\frac{\sin(nc\pi/2)}{n^2 \sin(c\pi/2)}$			
8A	Trapezoidal (no third harmonic)	$\sin\psi$	$c = 2/3$ in expressions on line 8		$\frac{12k}{\pi^2}\sin\frac{\pi}{3}$	1.053	$\frac{\sin(n\pi/3)}{n^2 \sin(\pi/3)}$	0.046	0.046	0.923
8B	Trapezoidal (no fifth harmonic)	$\sin\psi$	$c = 2/5$ in expressions on line 8		$\frac{20k}{\pi^2}\sin\frac{\pi}{5}$	1.191	$\frac{\sin(n\pi/5)}{n^2 \sin(\pi/5)}$	0.184	0.036	0.990

* Number before solidus is for $k = 1$, after solidus it is for $k \ll 1$.

waves. For convenience, these results are presented in Table 6.1 using the present nomenclature. In addition, two examples of trapezoidal waveforms and a case including only third harmonic are listed. A trapezoid selected to eliminate third harmonic would be best for single-phase output, while the case with no fifth harmonic would be useful for 3-phase output.

When only third harmonic is introduced in the control signal, pure sinusoidal output is recovered in a 3-phase system. The reference waves can be obtained by connecting 3-phase sinusoidal signals in wye and injecting a third harmonic (which need not be pure) between the neutral point and ground. This case is selected as an example for further analysis of a typical flat-topped waveform, because it is mathematically simple, yet closely approximates a trapezoid, which is probably more convenient in practice. It can be shown that the amount of third harmonic that results in a maximum ratio of fundamental to peak voltage is 1/6,

$$\bar{v}_0 = k \frac{2}{\sqrt{3}} \left(\sin \Omega t + \frac{1}{6} \sin 3\Omega t \right).$$

This waveform and its fundamental component are plotted in Figure 6.1; the figure also shows the trapezoid with no fifth harmonic.

A trapezoidal-wave generator is easily obtained by adding a clipping arrangement to a triangular-wave oscillator. For converters with 3-phase output, a 3-phase set of trapezoidal control signals can be obtained by means of a beat-frequency oscillator: Three flip flops that are driven in sequence at a constant frequency produce 3-phase quasi-square wave signals, which are then mixed with a variable frequency square wave. A set of three demodulators and low-pass filters now produce the desired difference-frequency trapezoidal signals.

The output waveforms produced by using linear ramp firing waves in conjunction with either triangular or sinusoidal control signals are also quite acceptable. The former combination is treated in detail in Section 5.4, while the latter case is analyzed by Clarke,[16] and the flat-topped wave obtained with full modulation ($k = 1$) is shown in Figure 6.2, together with its fundamental component. A trapezoid with no third harmonic is also shown.

For high output voltages, a flat-topped waveform can easily be produced by overmodulation, which drives the converter firing angle against the advance or retard limit stops. A sinusoidal control signal having $k > 1$ will produce an output voltage that is clipped at the

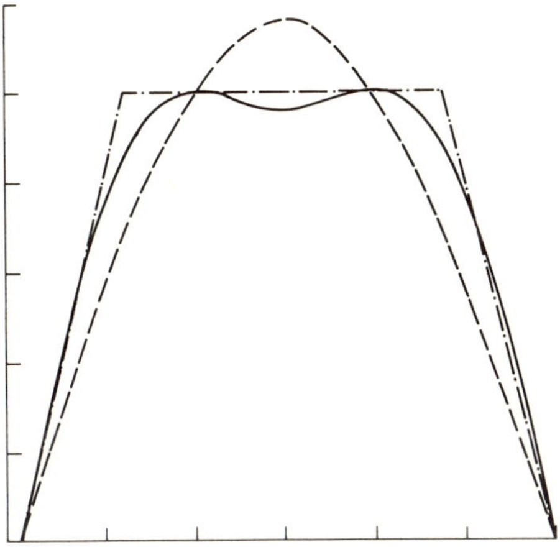

Figure 6.1. Output waveform with 16.7% third harmonic (solid line) and its fundamental component (dashed line). Trapezoid with no fifth harmonic (chain line).

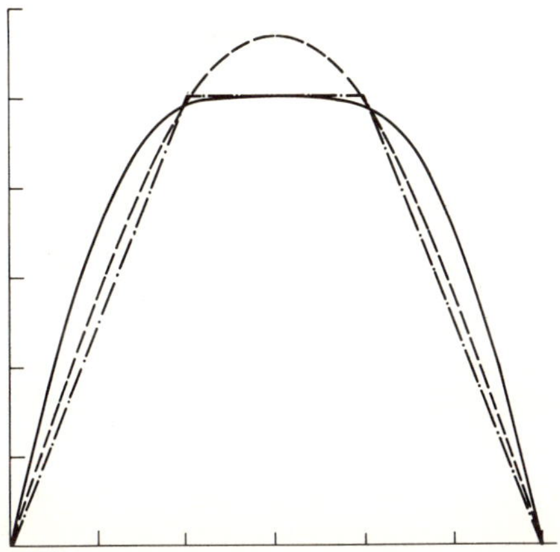

Figure 6.2. Output waveform obtained with linear firing waves and sinusoidal control signal for $k = 1$ (solid line) and its fundamental component (dashed line). Trapezoid with no third harmonic (chain line).

maximum value E_{do} when $\sin \Omega t > 1/k$. As k is increased, the waveform approaches a trapezoid. An overmodulated triangular control signal produces a true trapezoid.

However, overmodulation causes the gain to become nonlinear; the fundamental output voltage is no longer directly proportional to the control signal. The same defect is present in control methods 4, 5, and 6 that use linear firing waves. Where linear gain is important, a compensating nonlinearity could be introduced in the control circuit. Control methods 7 and 8 preserve a linear gain and generate the same flat-topped waveform at all voltage levels, although there is no advantage to this shape at low voltages.

Where sinusoidal output is a prime requirement, the conventional control method 1 is the obvious choice. When it is desired to improve the power factor and equipment utilization by programming a flat-topped waveform at high voltage levels, control method 4 or overmodulation of method 1 is suggested if linear gain is not critical. If linear gain is important, method 7 or 8 would be preferable. The performance achieved with triangular or square-wave signals (methods 2, 3, 5, and 6) is considered to be less than optimum.

In most practical applications, cycloconverters are operated under closed-loop feedback control, and the output voltage is not necessarily the controlled quantity. Motor drives, for example, often employ current regulators, and the voltage control signal v_0 is actually the current error signal. The voltage is automatically adjusted to meet the current demand. Since sinusoidal current is usually desired, the voltage also is approximately a sine wave. However, in a suitably connected 3-phase system, a flat-topped voltage wave is automatically generated at high output levels (see Section 11.3).

6.3 DISPLACEMENT POWER FACTOR

The pseudo-instantaneous Equations 4.17 and 4.19 for the real and reactive input power, respectively, can be written in more general terms that, through Equation 6.1, are applicable to the case of nonsinusoidal voltages.

$$\begin{aligned} P_r &= 0.5E_m A'_m \sin(\Omega t - \Phi) \sin\psi \\ &= 0.5E_m A'_m \sin(\Omega t - \Phi)\, \bar{v}_0 \, ; \end{aligned} \tag{6.4}$$

$$P_x = 0.5E_m A'_m \left| \sin(\Omega t - \Phi) \right| \sqrt{1 - \bar{v}_0^2}. \tag{6.5}$$

Thus, Equation 4.22 becomes

$$\frac{P_X}{P_R} = \frac{\int_{\Phi}^{\Phi+\pi} \sin(x - \Phi) \sqrt{1 - \bar{v}_0^2}\, dx}{\int_{\Phi}^{\Phi+\pi} \sin(x - \Phi)\, \bar{v}_0 \, dx}, \tag{6.6}$$

where $\bar{v}_0$ is a function of $x = \Omega t$, as given by Equation 6.2 or 6.3. The average displacement power factor P_1 follows, according to Equation 4.21. Computed values for P_1 as a function of k and cos Φ are presented in Figure 6.3 for the fourth control technique listed in Table 6.1, where the maximum value of P_1 (which is obtained when $k = \cos \Phi = 1$) for other techniques is included for comparison. Note that k in Figure 6.3 is not directly proportional to the fundamental output voltage, so that this figure is not directly comparable with Figure 4.6.

It must be stressed that these calculated values are for an ideal set of conditions: sinusoidal current, low frequency ratio, negligible overlap, and no margin for commutation when inverting with $k = 1$. The cases with square-wave signals are particularly unrealistic. When a known nonsinusoidal current is drawn by the load, the power factor

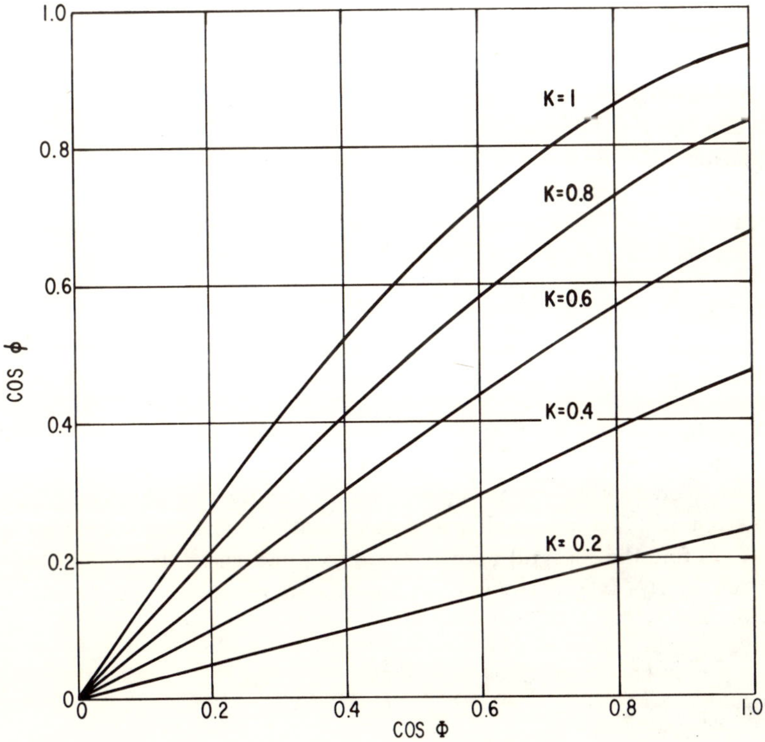

Figure 6.3. Variation of input power factor cos ϕ with output power factor cos Φ and modulation fraction k. Cycloconverter with linear firing waves and sinusoidal control signal.

can be computed by substituting the appropriate function in place of sin $(x - \Phi)$ in Equation 6.6. An example is the case of the delta-rectifier type of cycloconverter, described later in Section 9.4.

6.4 DISPLACEMENT POWER FACTOR CONSIDERING OVERLAP

When commutation overlap is considered, Equations 6.4 and 6.5 become

$$P_r = 0.5E_m A'_m \sin(\Omega t - \Phi)\, p_r$$

$$P_x = 0.5E_m A'_m \,|\sin(\Omega t - \Phi)|\, p_x,$$

where p_r and p_x are functions of α and u, given in Equations 2.36 and 2.37. If, as before, the range of integration is selected to be the conduction period of the positive current group, then α_p is the delay angle required and can be calculated at a given point $x = \Omega t$ from the programmed voltage at that point,

$$\alpha_p = \operatorname{Cos}^{-1} \bar{v}_0. \tag{6.7}$$

Now, knowing the instantaneous current a from Equation 4.9, the voltage v_x lost because of commutation overlap follows from Equation 5.27. Hence, the overlap angle u can be calculated from Equation 2.11 with $E_x = v_x$. Then p_r and p_x can be obtained from Equations 2.36 and 2.37 for use in the following modification of Equations 4.20 and 4.18:

$$P_X = \frac{E_m A'_m}{2\pi} \int_{\Phi}^{\Phi+\pi} \sin(x - \Phi)\, p_x \, dx; \tag{6.8}$$

$$P_R = \frac{E_m A'_m}{2\pi} \int_{\Phi}^{\Phi+\pi} \sin(x - \Phi)\, p_r \, dx. \tag{6.9}$$

The average displacement power factor can then be obtained from Equation 4.21, as before. If one notes Equations 2.1 and 4.13, the factor before the integral sign in Equations 6.8 and 6.9 can also be written as $E_{d0}A_m n_w/(sp\pi)$. Then the total input power to the converter is

$$\frac{P_R\, ps}{n_w} = E_{d0}A_m \frac{1}{\pi} \int_{\Phi}^{\Phi+\pi} \sin(x - \Phi)\, p_r \, dx. \tag{6.10}$$

This must, of course, equal the output power plus the thyristor and resistance losses,

$$\text{Input power} = VA \cos \Phi' + 0.9AU_a s + I_e^2 R_c \frac{ps}{n_w}. \qquad (6.11)$$

The calculation process outlined here and the solution of equations 6.8 and 6.9 may appear quite involved but are easily programmed on a computer. As verification, Equations 6.10 and 6.11 for the input power were found to be in agreement for a number of test cases. It is still assumed here that the frequency ratio is low. For computing the reactive input power P_X when the frequency ratio correction factors are included in the phasor diagram Figure 4.9, the phasor V_0 is taken in place of V_0. That is, the normalized programmed voltage $\overline{v}_0$ is multiplied by the ratio V_0''/V_0 before insertion in Equation 6.7, and the angle between the phasors V_0'' and A is used for Φ. It has been verified that this substitution results in the correct value of real input power, at least.

6.5 POWER FACTOR IMPROVEMENT BY UNSYMMETRICAL FIRING

With symmetrical firing, the displacement angle between the supply voltage and current is equal to the firing delay angle, neglecting the effect of commutating reactance. Thus, when a converter operates at low output voltage produced by a firing delay in the region about 90°, the power factor is low. Unsymmetrical firing can be used with converters that consist of two or more commutating groups connected in series. Essentially, one group is maintained at full advance (or full retard) so that its reactive power requirement is minimized, while the other group is controlled to produce the desired output voltage. This technique has also been called "sequential" or "consecutive" control.[9]

6.5.1 Principle of Operation

A more detailed description of unsymmetrical firing control will be given for the 3-phase bridge circuit, as an example. This relatively simple arrangement has disadvantages, as will be pointed out, but they can be overcome by using combinations of 3-phase bridges that would allow 12-pulse operation with the usual symmetrical firing. The following description pertains to ac/dc converters or to the pseudo-instantaneous operation of cycloconverters.

The 3-phase bridge converter is equivalent to two 3-pulse commutating groups connected in cascade with a $\pi/3$ radian displacement between their respective supply systems. It may be shown that the greatest improvement in power factor is obtained with the greatest asymmetry between the firing angles α_1 and α_2 of the two commutating groups, consistent with the required output voltage and firing angle limits. For maximum voltage (rectifying), both groups are fired at the advance limit α_0. To reduce voltage, one group (for example thyristors 1, 3, 5 in Figure 1.2) is held at the advance limit, while the other group (thyristors 2, 4, 6) is progressively phased back, until the retard limit α_r is reached.

At this point, the second group is producing a negative voltage opposing the positive voltage of the first group, with a resultant near-zero output voltage. Since the reactive power requirements of both a fully advanced rectifier and a fully retarded inverter are small, little reactive power is needed. The load current coasts straight through two thyristors in series (1 and 4, 3 and 6, or 5 and 2), bypassing the ac supply.

To reverse the net output voltage and operate as an inverter, the first group of thyristors is now phased back, while the second group is held at the retard limit α_r. When both groups fire at the retard limit, the maximum negative (inverting) voltage is produced. Thus, symmetrical firing occurs at each extreme of the operating range, which is therefore the same for both modes of firing.

In Figure 6.4, the cosine firing wave (A) that yields linear gain in a symmetrically controlled converter is compared with the offset cosine firing waveshapes (B) required to produce the same overall linear gain with unsymmetrical control of the two commutating groups. The wave in Figure 6.4(A) is the same for every thyristor in a given current polarity group and is repeated at 60° intervals for a 6-pulse converter (see Figure 1.3). The wave designated α_1 in Figure 6.4(B) repeats at the 3-pulse interval of 120° to control thyristors 1, 3, and 5. Similarly, the wave marked α_2 also repeats every 120° to control the 3-pulse group of thyristors 2, 4, and 6. However, the wave trains for the two groups have the mutual 60° displacement of their effective supply systems. Note that the firing angle retard limit α_r is usually less than $(180° - \alpha_0)$ to allow more margin for commutation in the inverting mode.

When unsymmetrical firing is applied to a 3-phase bridge converter, the following penalties are paid for the power factor improvement:

1. Third harmonic ripple is present in the output, as well as the usual sixth harmonic. However, the magnitudes of the third and sixth harmonics are less than for a simple 3-pulse converter.
2. Even harmonics are present in the supply line currents. This is

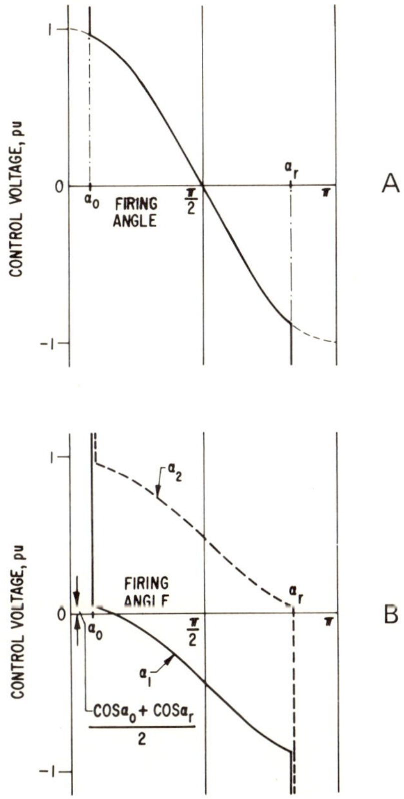

Figure 6.4. Firing waves for symmetrical and unsymmetrical control of a converter.
(A) Cosine wave for symmetrical firing.
(B) Offset cosine waves for unsymmetrical firing.

most undesirable for high-power equipment.

3. For two particular output voltage regions, there is interference between the commutations of the two commutating groups, through the common transformer and line reactance. The interference that occurs when one group is inverting at the retard limit can cause commutation failure.

4. The firing angle control circuitry is more complex.

For these reasons, unsymmetrical firing is not recommended for single 3-phase bridge converters. Unintended asymmetrical firing in the form of ripple instability is a familiar problem.

The drawbacks largely disappear when two 3-phase bridges are combined in the proper manner, in series or in parallel (with interphase transformers for the parallel case). Only the potential 12-pulse operation is sacrificed for the power factor improvement obtained with unsymmetrical firing. However, the sixth and twelfth harmonic components of output voltage are less than for a symmetrically fired 6-pulse converter.

The even harmonics in the line current are eliminated by the combination of two 3-phase bridges. In addition, since the supplies for the two bridges require separate transformer windings, it is easier to separate the commutating reactance and overcome penalty 3. This last statement assumes that each bridge is a symmetrical 6-pulse converter and they are operated unsymmetrically with respect to each other, with their outputs connected in series.

6.5.2 Calculation of Power Factor

With unsymmetrical firing delay angles α_1 and α_2, the output voltage contributions of each of the cascaded commutating groups are, still neglecting overlap, proportional to $\cos\alpha_1$ and $\cos\alpha_2$, respectively. The resultant per unit output voltage $\bar{v}_0$ is the algebraic sum of the separate contributions, and the pseudo-instantaneous real power coefficient p_r has the same value

$$\bar{v}_0 = p_r = \left(\frac{\cos\alpha_1 + \cos\alpha_2}{2}\right).$$

The per-unit reactive power p_x is, similarly, the algebraic sum of the reactive power requirements of the two commutating groups,

$$p_x = \left(\frac{\sin\alpha_1 + \sin\alpha_2}{2}\right).$$

Given the desired value of $\bar{v}_0$, the required firing angles and resulting pseudo-instantaneous reactive power coefficient p_x can be calculated as follows, neglecting overlap:
For

$$\cos\alpha_0 > \bar{v}_0 > \frac{(\cos\alpha_0 + \cos\alpha_r)}{2},$$

$$\alpha_1 = \alpha_0$$

$$\alpha_2 = \mathrm{Cos}^{-1}(2\bar{v}_0 - \cos\alpha_0)$$

$$p_x = \left[\frac{\sin\alpha_0 + \sqrt{1 - (2\bar{v}_0 - \cos\alpha_0)^2}}{2}\right]. \qquad (6.12)$$

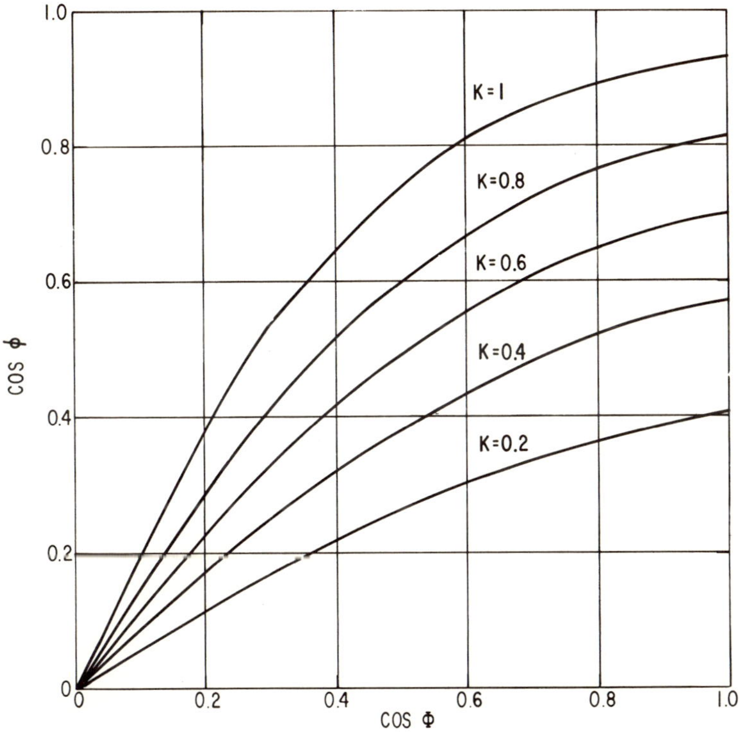

Figure 6.5. Variation of input power factor cos ϕ with output power factor cos Φ and modulation fraction k. Unsymmetrically fired cycloconverter with sinusoidal control signal, $\alpha_0 = 0$, $\alpha_r = \pi$.

For
$$\frac{(\cos\alpha_0 + \cos\alpha_r)}{2} > \bar{v}_0 > \cos\alpha_r,$$

$$\alpha_1 = \mathrm{Cos}^{-1}\,(2\bar{v}_0 - \cos\alpha_r)$$

$$\alpha_2 = \alpha_r$$

$$p_x = \left[\frac{\sin\alpha_r + \sqrt{1 - (2\bar{v}_0 - \cos\alpha_r)^2}}{2}\right]. \tag{6.13}$$

If $\alpha_0 = 0$ and $\alpha_r = \pi$, Equations 6.12 and 6.13 can be simplified and combined to give

$$p_x = \sqrt{|\bar{v}_0| - \bar{v}_0^2}.$$

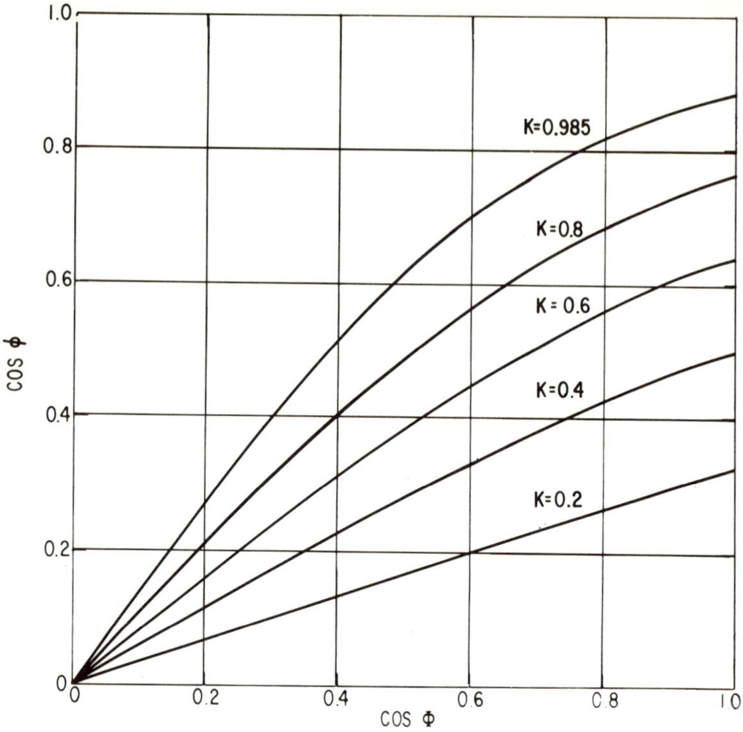

Figure 6.6. Variation of input power factor cos ϕ with output power factor cos Φ and modulation fraction k. Unsymmetrically fired cycloconverter with sinusoidal control signal, $\alpha_0 = 10°$, $\alpha_r = 150°$.

For symmetrical firing, the corresponding equation is

$$p_x = \sqrt{1 - \overline{v}_0^2}.$$

The ratio of the average reactive input power over one cycle of the output frequency to the average real power is given by Equation 6.6 with the appropriate substitution for the quantity under the radical sign. For a programmed sine-wave output, the denominator can be reduced to yield the following expression, and the average displacement power factor is again given by Equation 4.21:

$$\frac{P_X}{P_R} = \frac{\int_{\Phi}^{\Phi+\pi} \sin(x - \Phi)\, p_x \, dx}{0.5\pi k \cos \Phi}.$$

Computed values of the input power factor cos ϕ as a function of the output power factor cos Φ with the modulation fraction k as a parameter are plotted in Figures 6.5 and 6.6 for comparison with the case of symmetrical firing in Figure 4.6. The condition of unsymmetrical firing with the theoretical extreme firing angle limits $\alpha_0 = 0$ and $\alpha_r = 180°$ is presented in Figure 6.5 and shows a great improvement in power factor over Figure 4.6. For typical practical firing angle limits $\alpha_0 = 10°$ and $\alpha_r = 150°$, shown in Figure 6.6, the improvement is not so great, but it is still sufficient to make the use of unsymmetrical firing attractive. At 50% output voltage feeding a load of 0.8 power factor, the input power factor can be improved from 0.32 to 0.49. These values are both theoretical, but it is believed that the ratio of improvement should be maintained in practice.

Note that the maximum value of the modulation factor k is limited to $\cos \alpha_0$, or to $|\cos \alpha_r|$ when the load is regenerating. The limit $k = 0.985$ for $\alpha_0 = 10°$ is the top curve in Figure 6.6. These limits prevail in practice, where advance and retard limit stops must be employed, and are the same for symmetrical or unsymmetrical firing. In presenting the curves for $k = 1$ in Figure 4.6 and Figure 6.3, as well as for Figure 6.5, the theoretical extreme limits have been assumed possible.

7 Harmonics of the Supply Frequency in the Output of a Cycloconverter

In Chapter 5, distortion in the output of a cycloconverter caused by a high output/input frequency ratio was studied. Distortion introduced by some practical methods of controlling a converter was considered in Chapter 6. Both of these types of distortion produce harmonics of the output frequency and are not present in converters having dc output. In this section, distortion in the form of harmonics of the supply frequency will be analyzed. This is the familiar ripple voltage seen in the output of rectifiers.

For steady-state operation, the frequency spectrum of a rectifier consists of fixed harmonic voltage components occurring at multiples of the fundamental ripple frequency qf. The magnitudes depend upon the firing delay angle α and the overlap angle u and may be calculated by Fourier analysis of the output waveform, yielding the Fourier series coefficients quoted in Section 2.2. In a cycloconverter, the firing angle is modulated to produce the output voltage waveform, so that the harmonic content varies with a period equal to one half-cycle of the output. The frequency spectrum is no longer described by a simple series of harmonics but develops sidebands around each harmonic. In the general case where the frequency ratio is an irrational number, the output waveform is not repetitive, and its frequency spectrum is blurred into a continuous function, the Fourier transform of the actual time function.

The objective of this chapter is to develop a quantitative measure for the high frequency components in the output voltage of a cycloconverter as a function of the modulation fraction k and the pulse number q. Analysis of the general frequency spectrum appears to be a formidable task. However, if the output frequency is low compared with the input frequency, the harmonic coefficients for the rectifier case can be treated as pseudo-instantaneous quantities, just as the average dc voltage was treated as a pseudo-instantaneous variable in Section 4.1 Then, knowing the variation of the pseudo-instantaneous firing angle with time, the root-mean-square values of the harmonic coefficients over one output cycle can be calculated by integration. This yields a measure of the effective value of the true harmonics and their associated sidebands. Rather than attempt a derivation of explicit equations, a procedure for numerical integration by computer will be outlined and the results presented in graphical form. The validity of this approach diminishes as the frequency ratio increases.

For simplicity, the overlap angle u will be neglected, and only the principal harmonic of order q (the pulse number) will be considered. Also, the output frequency is assumed to be much lower than the supply frequency so that the pseudo-instantaneous principal harmonic coefficients are given by Equations 2.23 and 2.24 with $m = 1$. Expressing the ac crest values as fractions of the maximum direct voltage, one obtains

$$A_1 = \frac{a_1}{E_{do}} = \frac{\sin (q+1)\alpha}{q+1} - \frac{\sin (q-1)\alpha}{q-1}; \tag{7.1}$$

$$B_1 = \frac{b_1}{E_{do}} = \frac{\cos (q+1)\alpha}{q+1} - \frac{\cos (q-1)\alpha}{q-1}. \tag{7.2}$$

The values of these coefficients for the positive and negative current polarity groups are presented in Table 7.1 as functions of the control angle ψ defined in Equation 4.6, for pulse numbers $q = 2, 3, 4, 6$, and 12. Note that when q is an even number, then A_1 is the same for both groups while the B_1 coefficients have opposite signs. That is, the $(a_1 \sin q\,\omega t)$ terms of the Fourier series are in phase, while the $(b_1 \cos q\,\omega t)$ terms are 180° out of phase. Thus, in a cycloconverter with no blanking of one polarity group when the other is conducting, only the $(a_1 \sin q\,\omega t)$ component appears in the output, but the $(2b_1 \cos q\,\omega t)$ component appears across the intergroup reactor and drives the qth harmonic component of circulating current. Similarly, when q is an odd number, the component $(b_1 \cos q\,\omega t)$ is present in the output and $(2a_1 \sin q\,\omega t)$ produces circulating current.

Since the actual circulating current is pulsating dc, the qth harmonic component is not directly proportional to the total value but serves as a convenient comparative measure of the manner in which the circulating current varies with the firing angle and pulse number. Also, note that when a higher pulse number is obtained by cascading commutating groups having $p < q$, as in the 3-phase bridge circuit of Figure 1.2, then each commutating group has a circulating current of the lower order p. A detailed analysis of the circulating current for $p = 3$ is given by Bland.[57] In practical cycloconverters where no blanking logic is employed, the actual circulating current is usually reduced by differentially biasing the two polarity groups. The amount

Table 7.1 Coefficients of Sine and Cosine Components of Principal Harmonic of Supply Frequency in Cycloconverter Output.

Pulse No. q	Control Angle ψ (deg)	Positive Current Group A_1	Positive Current Group B_1	Negative Current Group A_1	Negative Current Group B_1
2	0	−1.333	0	−1.333	0
	15	−1.202	−0.495	−1.202	0.495
	30	−0.866	−0.833	−0.866	0.833
	45	−0.471	−0.943	−0.471	0.943
	60	−0.167	−0.866	−0.167	0.866
	75	−0.023	−0.73	−0.023	0.73
	90	0	−0.667	0	0.667
3	0	0	0.75	0	0.75
	15	−0.467	0.558	0.467	0.558
	30	−0.65	0.125	0.65	0.125
	45	−0.5	−0.25	0.5	−0.25
	60	−0.217	−0.375	0.217	−0.375
	75	−0.033	−0.308	0.033	−0.308
	90	0	−0.25	0	−0.25
4	0	0.533	0	0.533	0
	15	0.287	0.429	0.287	−0.429
	30	−0.173	0.433	−0.173	−0.433
	45	−0.377	0.094	−0.377	−0.094
	60	−0.233	−0.173	−0.233	0.173
	75	−0.043	−0.184	−0.043	0.184
	90	0	−0.133	0	0.133
6	0	−0.343	0	−0.343	0
	15	−0.015	−0.331	−0.015	0.331
	30	0.297	−0.029	0.297	0.029
	45	0.04	0.242	0.04	−0.242
	60	−0.171	0.049	−0.171	−0.049
	75	−0.055	−0.089	−0.055	0.089
	90	0	−0.057	0	0.057
12	0	0.168	0	0.168	0
	15	−0.162	0.004	−0.162	−0.004
	30	0.145	−0.007	0.145	0.007
	45	−0.119	0.01	−0.119	−0.01
	60	0.084	−0.012	0.084	0.012
	75	−0.043	0.014	−0.043	−0.014
	90	0	−0.014	0	0.014

of reduction is a trade off with the distortion introduced by the bias.

With blanking of the firing signals to the group of thyristors that are not carrying load current, the circulating current is suppressed, but the resultant harmonic voltage of magnitude

$$c_1 = \sqrt{a_1^2 + b_1^2}$$

appears in the output. This magnitude is the same for any control angle ψ, whether the positive or negative group is conducting. However, the phase changes when the load current reverses.

7.1 CONVERTER SUPPLYING SINGLE PHASE OR ISOLATED PHASE

The rms value of the harmonic coefficients can be obtained by the following numerical integration procedure, incorporated in a computer program that uses the Gauss's rule algorithm. An integration interval from 0 to $\pi/2$ radians of the output cycle is the necessary minimum. The integration routine selects a number of points $x_i = \Omega t_i$ within this interval, and the corresponding values α_i of the firing angle are obtained from Equation 4.4 or, for general inclusion of nonlinear operation, from

$$\alpha_{pi} = \mathrm{Cos}^{-1}\,[\bar{v}_0\,(x_i)]\,. \tag{7.3}$$

The corresponding firing angles of negative group thyristors are

$$\alpha_{ni} = \pi - \alpha_{pi}\,. \tag{7.4}$$

Since, in this case, the resulting rms harmonic does not depend upon which group is conducting the current, only the α_{pi} values need be used.

These values are then substituted in Equations 7.1 and 7.2 to yield coefficients A_{1i} and B_{1i}, which are squared, multiplied by a weighting factor dependent on i and summed for all the values of i. The square root of the sum divided by the integration interval gives the desired rms value. In the case of a converter with blanking, the process is a numerical solution of the equation

$$C_{1e} = \sqrt{\frac{2}{\pi}\int_0^{\pi/2} (A_1^2 + B_1^2)\,dx}\,.$$

For a converter with no blanking, the A_1^2 and B_1^2 terms are summed separately and selectively according to whether q is even or odd to give the components of qth harmonic in the output and across the intergroup reactor.

The computed results for sinusoidal output voltage are plotted against the modulation factor k in Figures 7.1 to 7.3. The numerical coefficients are the root-mean-square values of the pseudo-instantaneous crest values of the qth harmonic as a fraction of the maximum direct voltage E_{d0}. For sinusoidal output, they are equal to the rms values of the pseudo-instantaneous rms values of the qth harmonic as a fraction of the rms output voltage with $k = 1$. That is, for example,

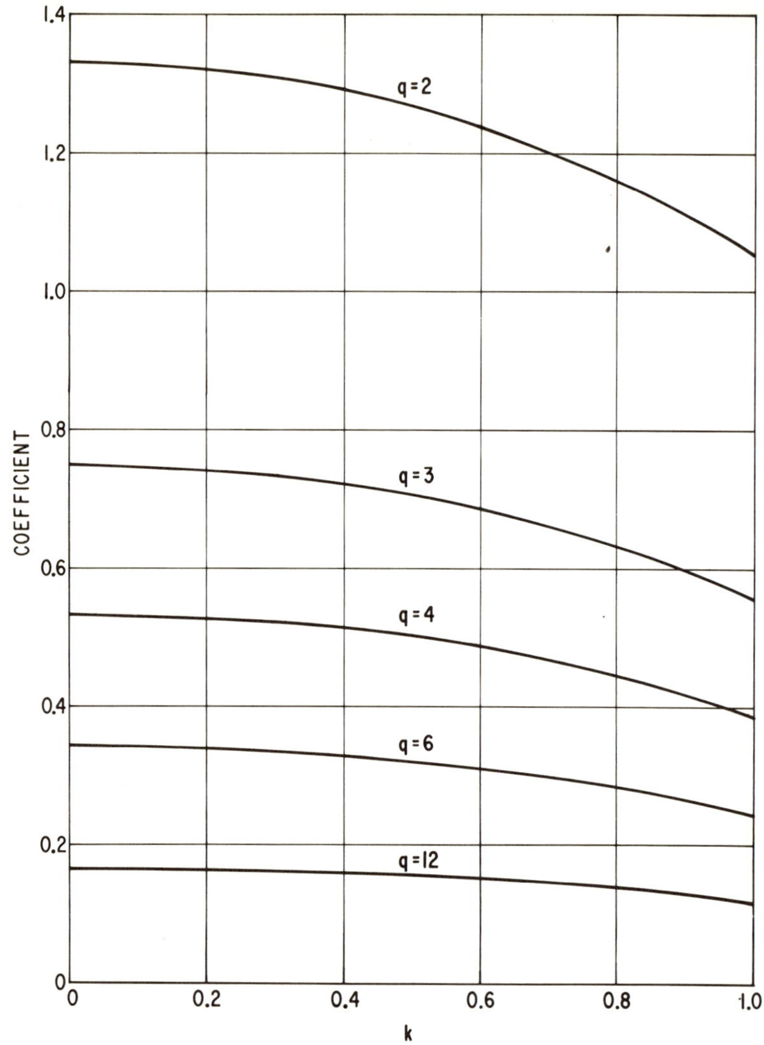

Figure 7.1. Coefficient of principal harmonic voltage in output of a cycloconverter with blanking, as a function of the modulation fraction k and pulse number q.

$$C_{1e} = \frac{c_{1e}}{E_{d0}} = \frac{c_{1e}/\sqrt{2}}{E_{d0}/\sqrt{2}},$$

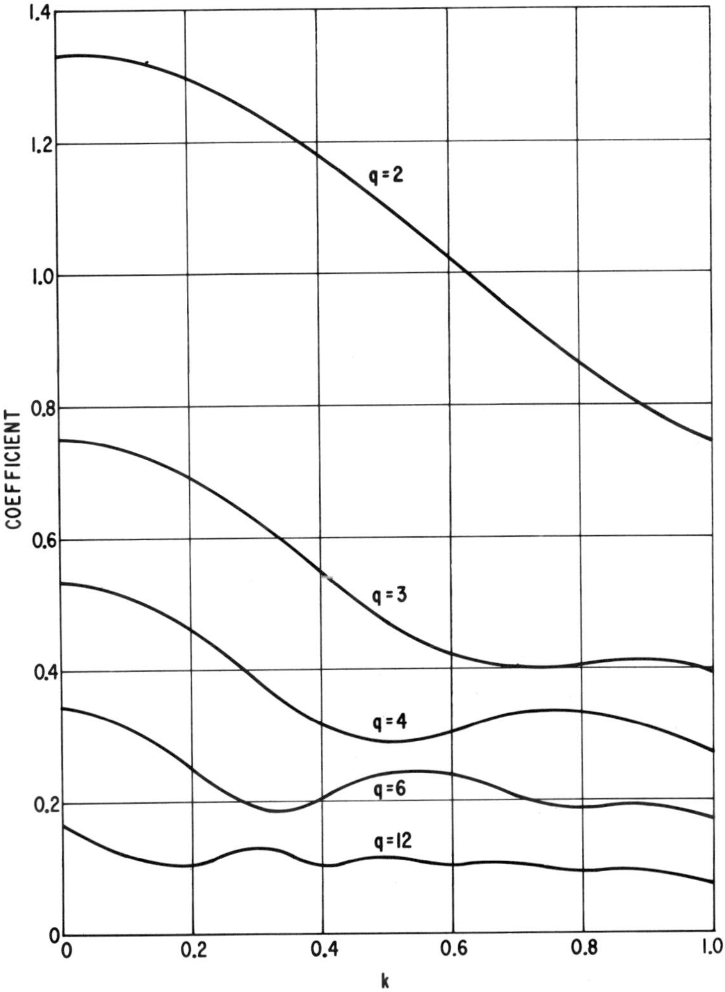

Figure 7.2. Coefficient of principal harmonic voltage in output of a cyclocon verter without blanking, as a function of the modulation fraction k and pulse number q.

where

$$c_{1e} = \sqrt{\frac{2}{\pi}\int_0^{\pi/2} c_1^2\, dx} = \sqrt{\frac{2}{\pi}\int_0^{\pi/2} (a_1^2 + b_1^2)\, dx.}$$

As expected, the harmonic content in the output becomes smaller as the modulation factor increases. The ratio of harmonic voltage with

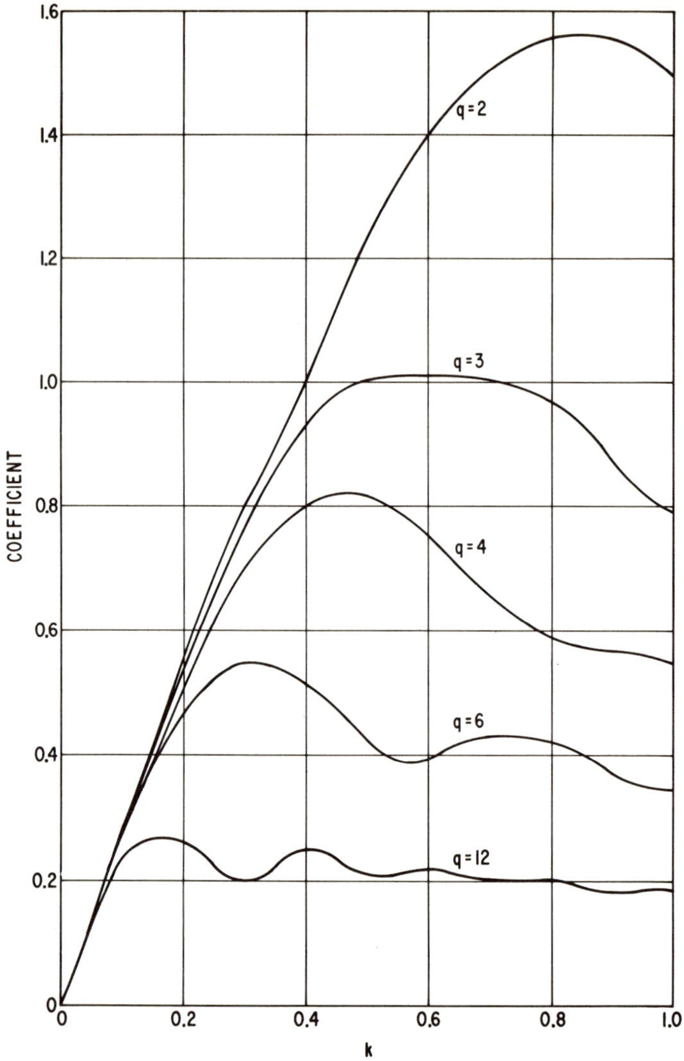

Figure 7.3. Coefficient of principal harmonic voltage across intergroup reactor of a cycloconverter without blanking, as a function of the modulation fraction k and pulse number q.

$k = 1$ to that with $k = 0$ is roughly 3/4 (with blanking) and 1/2 (with no blanking). In the case with no blanking, the intergroup component is a maximum at some intermediate value of modulation. This is confirmed by more exact analyses of the circulating current.[57] As in the case of ordinary rectifiers, the harmonic magnitude is reduced and its principal frequency is increased in proportion to the pulse number q, thus greatly easing the filtering requirements. Note that the curves in Figures 7.1 to 7.3 also give some of the higher order harmonics for converters having low pulse numbers, because q in Equations 7.1 and 7.2 is equivalent to mq in Equations 2.23 and 2.24. For example, the curves for $q = 4, 6,$ and 12 give the harmonics of order $m = 2, 3,$ and 6, respectively, in a 2-pulse converter, and the curves for $q = 6$ and 12 give the harmonics of order $m = 2$ and 4 in a 3-pulse converter.

The output of a fully modulated 3-pulse cycloconverter operated without circulating current will contain about 56% rms "ripple" voltage at the third harmonic of the supply frequency, 24% at the sixth harmonic, and 12% at the twelfth harmonic, as obtained from Figure 7.1. For a 6-pulse converter, the third harmonic will be eliminated, but the other values remain the same. The ripple from a cycloconverter is considerably higher than for a fully advanced rectifier or a fully retarded inverter. However, for an output voltage of zero ($k = 0$, $\alpha =$ 90°) both types of converter have the same harmonic output.

7.2 CONVERTERS CONNECTED IN WYE WITH ISOLATED NEUTRAL

If three similar converters having the same supply phases are connected in wye to give 3-phase output, the net pseudo-instantaneous harmonic content between pairs of output lines is the difference between the component produced by a phase firing at angles given by Equations 7.3 and 7.4 and the component produced by one of the other two phases firing at angles given by

$$\alpha'_{pi} = \mathrm{Cos}^{-1}\left[v_0\left(x_i \pm \frac{2\pi}{3}\right)\right]; \qquad (7.5)$$

$$\alpha'_{ni} = \pi - \alpha'_{pi}.$$

It has been recognized that there is some harmonic cancellation because of the wye connection, but the amount has not previously been determined. The computer program was adapted to calculate the net rms harmonic content of the wye connection.

The same result obtains from selecting either the + (leading) or − (lagging) phase in Equation 7.5. For converters with blanking, the resultant harmonic component depends upon the power factor of the load, making an integration interval of π output radians necessary. Firing angles α_{pi} or α_{ni} are selected according to whether the current in the reference output phase at time point x_i is positive or negative. Similarly, firing angles α'_{pi} or α'_{ni} are selected according to the instantaneous polarity of current in the displaced output phase at the same time. The firing angles thus chosen are substituted in Equations 7.1 and 7.2 to obtain corresponding values of the harmonic coefficients. The program then evaluates the following integral numerically

$$C_{1e} = \frac{1}{\sqrt{3}}\sqrt{\frac{1}{\pi}\int_0^{\pi}\left[(A_1 - A_1')^2 + (B_1 - B_1')^2\right] dx}, \tag{7.6}$$

where the factor $1/\sqrt{3}$ converts the line-to-line value to the equivalent line-to-neutral value, so that C_{1e} is the rms – rms value of the qth harmonic as a fraction of the rms output phase voltage with $k = 1$.

For a converter with no blanking, the voltage waveform theoretically does not change with the power factor of the load. The net harmonic content is obtained by evaluating Equation 7.6 with either the term $(A_1 - A_1')$ or $(B_1 - B_1')$ eliminated, depending upon whether q is odd or even.

The computed harmonic content with purely resistive or reactive load for a converter with blanking, and with any load for a converter with no blanking, are plotted in Figures 7.4 and 7.5 for comparison with Figures 7.1 and 7.2, respectively. Note that the harmonic cancellation is generally appreciable, but it varies in a complex manner dependent upon the pulse number q and the modulation factor k. For $k = 0$, the cancellation is complete, because all thyristors fire at $\alpha = \pi/2$ at all times, thereby producing the same output voltage waveshape in all three phases; as a result, the difference between phases is zero.

For a converter with blanking, the harmonic content is somewhat less when the load is resistive. For the practically most interesting case of a 3-phase wye connection of converters having $p = q = 3$ (Figure 1.1 circuit), the computer program was modified to print Table 7.2, detailing the variation of harmonic content with load power factor. It is seen that most of the change occurs between a power factor of 0.8 and 1.0.

From Figure 7.4, it can be seen that the output of a fully modulated 3-pulse converter, operated without circulating current and with a 3-phase wye-connected load of unity power factor will contain about 45% rms "ripple" voltage at the third harmonic of the supply frequency, 19% at the sixth harmonic, and 9% at the twelfth harmonic. Compared with the same converter without the wye output connection,

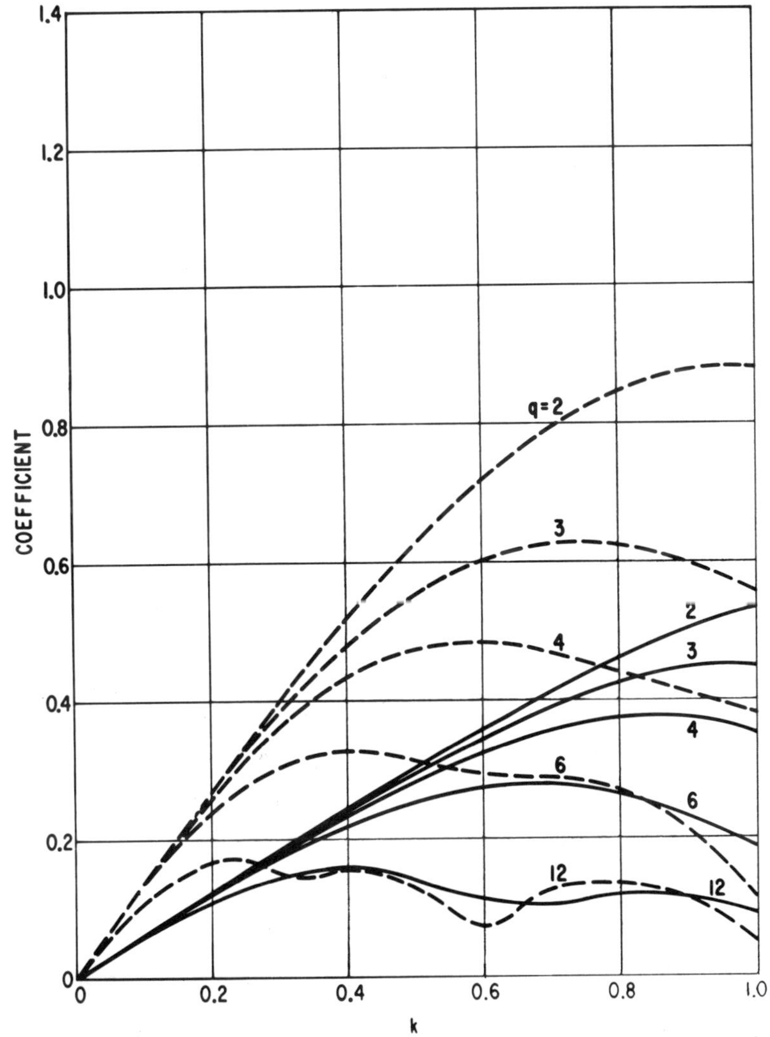

Figure 7.4. Coefficient of principal harmonic voltage in output of 3-phase wye-connected cycloconverters with blanking, as a function of the modulation fraction k and pulse number q. Solid lines : pure resistive load. Dashed lines : pure reactive load.

which was presented as an example in the last paragraph of Section 7.1, the harmonic content is about 20% lower. With low modulation, the improvement is much greater.

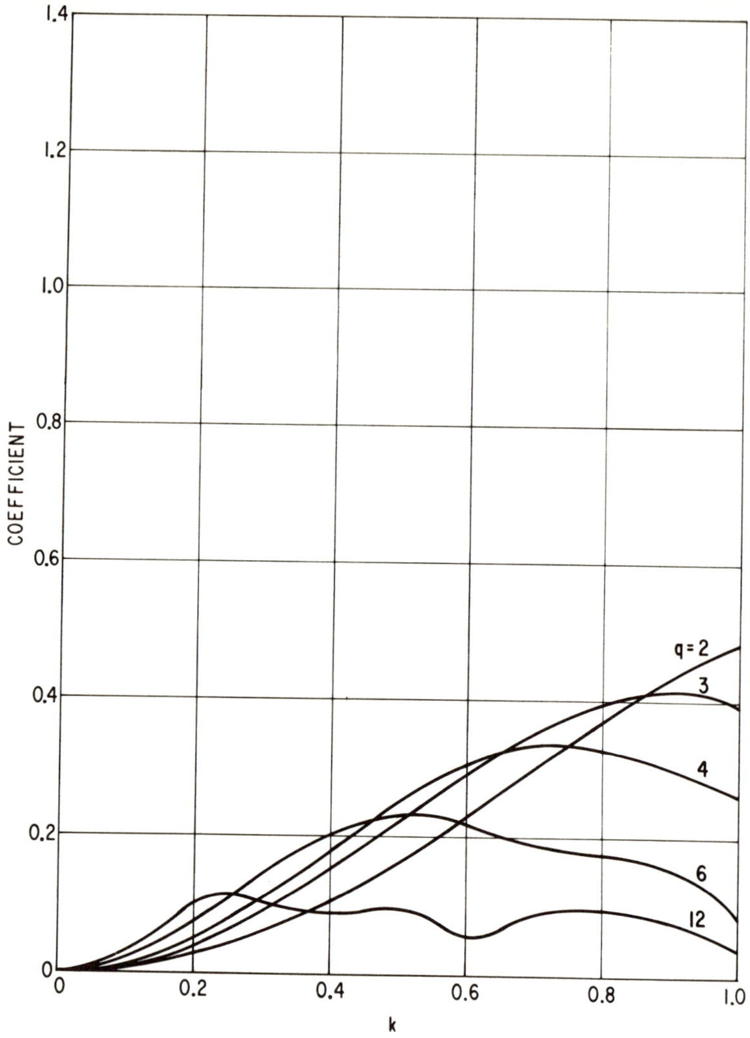

Figure 7.5. Coefficient of principal harmonic voltage in output of 3-phase wye-connected cycloconverters without blanking, as a function of the modulation fraction k and pulse number q.

Since the wye connection of converters also benefits from nonlinear operation, an example for the case of a linear firing wave combined with a sinusoidal signal is shown in Table 7.3. This method of operation is Control Method 4 in Table 6.1 and with $k = 1$ produces the

Table 7.2. Principal Harmonic of Supply Frequency in Cycloconverter Output. 3-Phase Output Wye Connected. Pulse No. $q = 3$. (Table gives per unit qth harmonic in output.)

Load Power Factor Modulation Factor k	0	0.2	0.4	0.6	0.8	1
0	0	0	0	0	0	0
0.1	0.135	0.133	0.126	0.114	0.094	0.061
0.2	0.263	0.26	0.247	0.225	0.186	0.122
0.3	0.379	0.375	0.357	0.326	0.272	0.181
0.4	0.478	0.473	0.452	0.416	0.35	0.238
0.5	0.554	0.548	0.527	0.489	0.418	0.292
0.6	0.604	0.599	0.58	0.544	0.473	0.342
0.7	0.627	0.624	0.609	0.578	0.513	0.386
0.8	0.624	0.622	0.613	0.591	0.537	0.422
0.9	0.599	0.598	0.595	0.582	0.541	0.447
1.	0.559	0.559	0.559	0.554	0.525	0.45

Table 7.3. Principal Harmonic of Supply Frequency in Cycloconverter Output. 3-Phase Output Wye Connected. Linear Firing Wave with Sinusoidal Signal. Pulse No. $q = 3$. (Table gives per unit qth harmonic in output.)

Load Power Factor Modulation Factor k	0	0.2	0.4	0.6	0.8	1
0	0	0	0	0	0	0
0.1	0.209	0.206	0.195	0.178	0.147	0.095
0.2	0.391	0.386	0.368	0.336	0.28	0.186
0.3	0.525	0.52	0.499	0.461	0.391	0.266
0.4	0.604	0.599	0.58	0.544	0.471	0.333
0.5	0.629	0.626	0.613	0.585	0.519	0.383
0.6	0.615	0.614	0.608	0.59	0.537	0.415
0.7	0.579	0.579	0.579	0.57	0.53	0.429
0.8	0.537	0.537	0.538	0.535	0.505	0.426
0.9	0.494	0.494	0.494	0.493	0.472	0.412
1.	0.451	0.451	0.451	0.451	0.437	0.393

flat-topped waveform shown in Figure 6.2. Comparing Tables 7.2 and 7.3 at the maximum power output point $k = 1$, $\cos \Phi = 1$, the harmonic coefficients are 0.45 and 0.393, respectively. However, the flat-topped wave has a fundamental component 1.133 per unit greater than a sine wave of the same crest value. Thus, the per-unit harmonic coefficient based on the fundamental output is reduced to 0.393/1.133 = 0.347, or 77% of the 0.45 per unit value for sinusoidal operation.

Although the difference is not outstanding, it is another advantage of the wye connection combined with a flat-topped voltage waveform to be considered in addition to the improved input power factor and reduced thyristor voltage. In the case of single-way converters (notably, 3-pulse), a 3-terminal load can be supplied without isolating transformers. Thus, the arrangement has optimum utilization of components and is recommended whenever the frequency ratio limit criterion (see Figure 5.10) permits its use.

As a reminder, it should be understood that the "harmonic" voltage values calculated in this section and presented in the figures and tables are not true harmonics according to the strict definition of the term. Each component in the discrete frequency spectrum of a rectifier, which is truly harmonic, generates a family of sideband frequencies when cycloconverter modulation is applied. The "harmonic" quantities of this section are suggested as practical measures of the total effect of each family. As the frequency ratio increases, the principal members of the families diverge in frequency from the true harmonic, and members of different families overlap. Problems with low subharmonic "beat" frequencies can arise. A detailed analysis of the output frequency spectrum is given by Pelly.[9]

8 Thyristor Voltage Waveform and Associated Losses

The waveform of voltage appearing across an individual thyristor in a converter circuit is of interest to the designer because of the sudden step changes that occur. The rate of voltage rise (dv/dt) must be limited so that the charging current of the junction capacitance does not trigger the device into conduction. This is accomplished by connecting a "snubber," comprising of a capacitor in series with a damping resistor, across each device. The snubber acts in conjunction with the commutating inductance that is effectively in series with the device. Snubbers also absorb the energy associated with the reverse recovery current of the thyristors, which is trapped in the commutating inductance at the time when they block.

It is shown in Section 8.1 that, for nonreversing converters, positive voltage steps appear in 3-phase bridge circuits but not in other common connections. However, in reversing converters, particularly cycloconverters, the negative steps on one polarity group appear as positive voltage steps across the other group of thyristors. The design of snubber circuits is discussed in Section 8.2.

When the supply frequency is 60 Hz or less, the average losses in the snubbers are generally insignificant compared with the power output, although the resistors must be sized to absorb a high peak energy impulse. However, cycloconverters used for power conditioning may be fed from high-frequency generators at several kilo-Hertz. Here, the average snubber loss may become appreciable and reduce the efficiency of the equipment. Equations for calculating the snubber losses are derived in Section 8.3.

The size and loss of the snubbers can be reduced by using thyristors having higher dv/dt capability. In Section 8.4, a relationship between the voltage regulation of the converter, its per unit snubber loss, and the dv/dt rating of the component thyristors is derived. Thyristor turn-on losses resulting from discharge of the snubbers are also studied. An expression for the rms thyristor voltage is given in Section 8.5, allowing calculation of the loss in shunt-connected resistors used for voltage equilization in series strings of devices.

8.1 STEPS IN THE THYRISTOR VOLTAGE WAVEFORM

Step changes in the voltage across a thyristor in a rectifier circuit occur whenever the dc line or the ac line between which it is connected is involved in a commutation. A dc line is involved in p commutations per cycle; a particular thyristor takes part in two of these, one at turn on that produces a negative step of $E_{cm}\sin\alpha$, the second at turn off giving a negative step of $E_{cm}\sin(\alpha + u)$. The $p - 2$ commutations between other thyristors each produce a pair of negative steps across our particular thyristor, $0.5E_{cm}\sin\alpha$ at the beginning of a commutation and $0.5E_{cm}\sin(\alpha + u)$ at the end. In single-way (star) circuits, these are the only steps.

In double-way (bridge) circuits, the ac line is shared with another thyristor, connected to the opposite dc line, which produces a notch at the common ac point during its incoming commutation and another notch during its outgoing commutation. Across our particular thyristor, these two commutation notches produce a positive step of $0.5E_{cm}\sin\alpha$ at a phase angle $(\alpha + \pi)$, measured from the point of natural commutation, followed after the overlap angle u by a negative step of $0.5E_{cm}\sin(\alpha + u)$, and after an angle $2\pi/p$ by a second negative step of $0.5E_{cm}\sin\alpha$ followed by a second positive step of $0.5E_{cm}\sin(\alpha + u)$ at a phase angle $(\alpha + \pi + 2\pi/p + u)$. The sizes of the voltage steps given in this paragraph are for the case when all the commutating inductance is in the ac lines and none is in the individual legs. When p is an odd number, all these commutations are distinct, and the described steps appear in the thyristor voltage waveform as illustrated in Figures 8.1 and 8.2 for a 3-phase bridge operating as a rectifier and inverter, respectively, under the same conditions depicted in Figure 2.1(A) and 2.1(B).

However, in double-way circuits having an even number of phases p, the notches caused by the thyristor between the same ac phase and the opposite dc line are simultaneous with the commutation steps produced by the thyristor between the same dc line and the 180° opposite phase. As seen by our particular thyristor, the positive steps due to the ac line notches are neutralized because the dc line steps in the same direction, while the negative steps due to the ac line notches are doubled because the dc line steps in the opposite direction, with respect to ground. The single-phase ($p = 2$) bridge circuit is an exception to this general argument.

It is seen that positive steps in thyristor voltage occur only in odd-phase bridge circuits, of which the 3-phase bridge is the only common example. If the thyristor is forward biased, such positive steps may

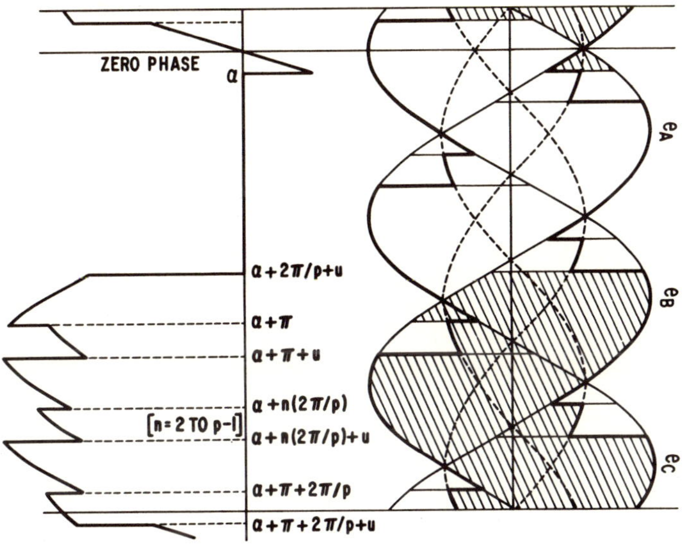

Figure 8.1. Thyristor voltage waveform in 3-phase bridge rectifier. (Shaded area between supply voltages at top.)

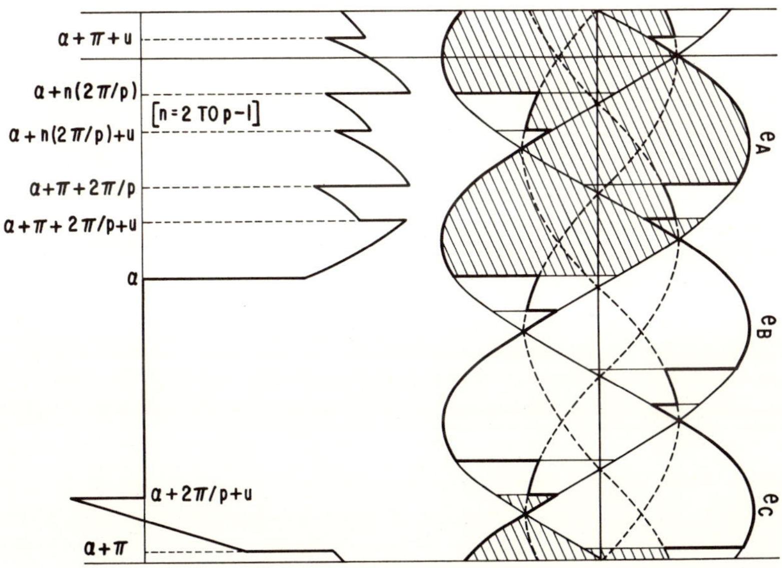

Figure 8.2. Thyristor voltage waveform in 3-phase bridge inverter. (Shaded area between supply voltages at top).

cause dv/dt firing if the slope is not adequately limited by the snubber circuits.

8.2 DESIGN OF SNUBBER CIRCUITS

In cycloconverters and reversing rectifiers, all the negative steps of one polarity group appear as positive steps across the thyristors in the opposite polarity group, if the two groups are not completely decoupled. If thyristors are connected directly in inverse parallel (no intergroup reactors), there is, of course, no decoupling, and the common snubber circuits must be sized to limit the reverse dv/dt across a recovering thyristor to the safe forward dv/dt capability of its mate. Otherwise, the reverse voltage spike would be the only concern.

The snubber elements comprise an effective inductance L_e in series with a thyristor, and a capacitor C_d in series with a damping resistor R_d connected across the thyristor. If the reverse recovery current of the thyristor is negligible, it may be shown than an optimum compromise between the peak transient voltage and the average dv/dt slope to that peak, in the sense of minimizing their product for a given L_e and C_d, is achieved when the damping factor $\zeta = 0.964$. The peak voltage is, then, $1.14E_c$, and the average slope to this value is

$$\left(\frac{dv}{dt}\right)_{\text{avg}} = 0.564 \frac{E_c}{\sqrt{L_e C_d}}. \tag{8.1}$$

If the inductance L_e effective at the time of current zero is taken to be $2L_c + L_u$, then for a specified dv/dt that will not trigger the reverse thyristor with the maximum value E_{cm} of the commutating voltage E_c, the necessary snubber capacitance C_d is given by

$$C_d = \frac{1}{2L_c + L_u}\left[\frac{0.564E_{cm}}{(dv/dt)_{\text{avg}}}\right]^2. \tag{8.2}$$

The optimum damping resistance is

$$R_d = 2\zeta\sqrt{\frac{L_e}{C_d}} = \frac{2 \times 0.964\,(2L_c + L_u)\,(dv/dt)_{\text{avg}}}{0.564E_{cm}}.$$

Even if the forward and reverse polarity groups are separated by intergroup reactors so that the value of dv/dt in Equation 8.1 can be greater than the safe forward dv/dt, it is generally found that snubbers of about the same size are required to suppress the reverse voltage spike caused by the reverse recovery current of the thyristor.

8.3 SNUBBER LOSSES

Every time that a voltage step of magnitude U_s occurs in the voltage across a thyristor, an amount of energy $0.5C_dU_s^2$ is lost in the associated snubber resistance R_d. As discussed in Section 8.1, let N_1 be the number of pairs of "full steps" of size $E_{cm} \sin \alpha$ and $E_{cm} \sin (\alpha + u)$, and let N_2 be the number of pairs of "half steps" of size $0.5E_{cm} \sin \alpha$ and $0.5E_{cm} \sin (\alpha + u)$. Values of N_1 and N_2 are presented in Table 8.1 for various circuit configurations. The losses that occur during the smoothly varying portions of the waveform, such as a sinusoidal supply voltage, are generally negligible in comparison with the step losses.

The loss in each snubber is

$$\begin{aligned} w_s &= 0.5fC_dE_{cm}^2[N_1 \sin^2\alpha + N_1 \sin^2 (\alpha + u) + 0.25\, N_2^2 \sin^2 \alpha \\ &\qquad + 0.25\, N_2^2 \sin^2 (\alpha + u)] \\ &= 0.5fC_dE_{cm}^2 \left(N_1 + \frac{N_2}{4}\right) [\sin^2 \alpha + \sin^2 (\alpha + u)] . \end{aligned}$$

If the overlap is small so that $(\alpha + u) \sim \alpha$ but not negligible such that some steps merge, then

$$w_s \approx fC_dE_{cm}^2 \left(N_1 + \frac{N_2}{4}\right) \sin^2 \alpha, \tag{8.3}$$

and the maximum loss occurs when $\sin \alpha = 1$, or $\alpha = \pi/2$. The total snubber loss is equal to the total number of thyristors times Equation 8.3, or half that number in the case of inverse-parallel pairs sharing common snubbers. In this latter case, the maximum total snubber loss is (for $\alpha = \pi/2$ at all times)

$$W_{sm} = n_0 psfC_dE_{cm}^2 \left(N_1 + \frac{N_2}{4}\right). \tag{8.4}$$

Now, in a sinusoidally modulated cycloconverter (see Section 4.1), the mean value of $\cos^2\alpha$ is $k^2/2$, so that the mean value of $\sin^2\alpha$ is $(1 - k^2/2)$ and the average snubber loss is

$$W_s = W_{sm}\left(1 - \frac{k^2}{2}\right), \tag{8.5}$$

which is a minimum of $0.5W_{sm}$ when $k = 1$.

If there are snubbers C'_d, R'_d connected in polygon between the p

Table 8.1. Numbers of Voltage Step Pairs in Common Converter Circuits.

Phases, p	Ways*	N_1	N_2
2	1	1	0
	2	1	0
3	1	1	1
	2	1	3
4	1	1	2
	2	2	0
5	1	1	3
	2	1	5
6	1	1	4
	2	2	2

* Single way (star) = 1; double way (bridge) = 2

phases of the ac supply system, their losses can be estimated as follows: the normal voltage across each snubber follows the sinusoidal commutating voltage e_c. For a single-way connection, there is a "full step" voltage notch caused by the commutation between the two phases spanned by the snubber. One of these phases is involved in the previous commutation, and the other phase is involved in the next commutation, which, respectively, produce a "half-step" notch and a "half-step" spike in the phase-to-phase voltage. In bridge circuits, the second commutating group causes a similar number of voltage steps across the same snubber.

Thus, the loss in each phase-to-phase snubber is similar to Equation 8.3 with $N_1 = n_w$ and $N_2 = 2n_w$, or $(N_1 + N_2/4) = 1.5n_w$. Multiplying by the number of snubbers (n_0ps/n_w), one obtains the total loss,

$$W_s' = 1.5n_0psfC_d'E_{cm}^2\left(1 - \frac{k^2}{2}\right).$$

The bi-phase commutating group $(p = 2)$ is an exception to the general argument. It has one phase-to-phase snubber, which is subjected to two "full-step" notches per cycle, so that $N_1 = 2$, $N_2 = 0$. In a complete cycloconverter, there are (n_0s/n_w) such snubbers, and their total loss is

$$W_s' = 2n_0\left(\frac{s}{n_w}\right) fC_d'E_{cm}^2\left(1-\frac{k^2}{2}\right).$$

8.4 LOSSES AND THYRISTOR *dv/dt* RATING

The snubber losses can become a significant fraction of the total converter losses when the supply frequency is high, and thyristors having a high *dv/dt* capability are desirable. A relationship between the relative snubber loss and the *dv/dt* rating of the thyristors can be derived as follows, assuming snubbers across the thyristors only: Substituting Equation 8.4 into 8.5 with the value of C_d given by Equation 8.2, one obtains

$$W_s = \frac{(0.564)^2\, n_0 psf E_{cm}^4\, (N_1 + N_2/4)(1 - k^2/2)}{(2L_c + L_u)\,(dv/dt)_{\text{avg}}^2}. \tag{8.6}$$

The snubber loss as a fraction of the output power with full modulation will be termed the per-unit snubber loss σ, and is obtained by dividing Equation 8.6 by 4.32 with $k = 1$. The voltage E_{do} is expressed in terms of E_{cm} through Equations 2.1 and 2.26,

$$E_{do} = \frac{E_{cm} sp}{2\pi}$$

$$\sigma = \frac{W_s}{W_0} = \left[\frac{0.564 E_{cm}\omega}{(dv/dt)_{\text{avg}}}\right]^2 \cdot \frac{(N_1 + N_2/4)}{(1-\epsilon)\,\epsilon} \cdot \frac{L_c}{2L_c + L_u}. \tag{8.7}$$

The designer may select the tolerable per-unit snubber loss σ, and the *dv/dt* required of the thyristors follows by rearrangement of Equation 8.7. If saturating reactors are not desired, $L_u = 0$ and

$$\left(\frac{dv}{dt}\right)_{\text{avg}} = 0.564 E_{cm}\omega \sqrt{\frac{N_1 + N_2/4}{2\sigma(1-\epsilon)\,\epsilon}}.$$

For a 3-phase bridge circuit, $p = 3$, $N_1 = 1$, $N_2 = 3$, and E_{cm} is equal to the peak thyristor voltage U_m. Then,

$$\left(\frac{dv}{dt}\right)_{\text{avg}} = \frac{0.564\,\pi\sqrt{3.5}\, fU_m}{\sqrt{\sigma(1-\epsilon)\,\epsilon}} = \frac{3.31\, fU_m}{\sqrt{\sigma(1-\epsilon)\,\epsilon}}. \tag{8.8}$$

Thus, for example, if thyristors operate at 1000 V (peak) and 2 kHz in a cycloconverter for which a regulation of 5% and a tolerable snubber loss of 2% are specified, their required *dv/dt* is

$$\left(\frac{dv}{dt}\right)_{avg} = \frac{3.31 \times 2000 \times 1000}{\sqrt{0.02 \times 0.95 \times 0.05}} \times 10^{-6} \text{V}/\mu\text{sec}$$
$$= 215 \text{ V}/\mu\text{sec}.$$

When the reverse recovery current of the thyristors is appreciable under the prevailing current and *di/dt* conditions, the factor 0.564 in Equation 8.8 will be greater, so that higher *dv/dt* will be necessary if the losses are not to increase.

Because of the turn-on time of the thyristor, some of the snubber discharge energy will be dissipated in the thyristor instead of the damping resistor, and the capacitor discharge increases the initial *di/dt* in the thyristor. An estimate of the fraction of the snubber discharge loss that is absorbed by the thyristor can be derived if turn on is postulated to be an exponential voltage fall with time constant τ. A simple analysis shows that the loss is divided between the thyristor and its snubber in proportion to their respective time constants. The ratio of the thyristor switching loss caused by snubber discharge to the loss produced by the rising load current is approximately

$$\frac{W_t}{W_L} \approx \frac{L_e/R_d}{\tau} = \frac{0.293E_{cm}}{\tau(dv/dt)_{avg}}.$$

If the fall time is in the order of $1\,\mu$sec, $E_{cm} = 1000$ V and $dv/dt =$ 215 V/μsec, as in the previous example, then the snubber discharge dissipation is about 135% of the load *di/dt* loss.

Thus, much of the demand for high *di/dt* in cycloconverters is a result of the need to discharge the snubber capacitors and can be relieved by increasing the *dv/dt* capability. In most inverters and choppers, the snubber circuits can be polarized by using auxiliary diodes and resistors to make the discharge time constant much longer than the fall time so that the thyristor dissipation is reduced. However, in cycloconverters, the snubber is often common to an inverse-parallel pair of thyristors and must have the same time constant for both voltage polarities.

8.5 VOLTAGE EQUALIZER LOSSES

When thyristors are connected in series for high-voltage operation, voltage equalizing resistors are usually connected across each device. If U_e is the total rms voltage across a series string of n_s thyristors and equalizing resistors of R_s ohms each, then the loss in each resistor is, neglecting any residual voltage unbalance,

$$W_r = \frac{U_e^2}{n_s^2 R_s}. \tag{8.9}$$

It can be shown that the rms thyristor voltage U_e in a phase-controlled rectifier, neglecting overlap, is

$$U_e = E_m \sqrt{1 + \frac{p}{4\pi} \sin\frac{2\pi}{p} (2\cos^2\alpha - 1)},$$

which is a maximum when $\alpha = 0$ and a minimum when $\alpha = \pi/2$.

For a cycloconverter with sinusoidal output modulation, the mean value of $\cos^2\alpha$ is $k^2/2$; therefore

$$U_e = E_m \sqrt{1 - (1 - k^2)\frac{p}{4\pi} \sin\frac{2\pi}{p}}. \tag{8.10}$$

Equation 8.10 can be substituted into Equation 8.9 to give the loss W_r in each resistor, which is a maximum when $k = 1$,

$$W_{r\,\max} = \frac{E_m^2}{n_s^2 R_s}.$$

This should be multiplied by a factor to allow for the worst-case unbalance before determining the resistor rating. The total equalizing resistor loss for the complete circuit is, of course, the total number of thyristors ($2n_0psn_s$) times W_r, or half this value if inverse-parallel pairs of thyristors share a common equalizing resistor. Transient voltage equalization usually requires an increase in the size of the snubber circuits, which increases their dissipation, of course.

The blocking losses of the thyristors can be treated in a similar manner, assuming that their leakage can be represented by an equivalent resistance. However, a better representation may be an assumed constant leakage current, independent of voltage, in which case the thyristor blocking loss is approximately that current times U_e, Equation 8.10.

9 The Delta-Rectifier Cycloconverter

In this circuit configuration, there are no distinct forward and reverse current polarity groups of thyristors. Instead, three unipolarity rectifying groups are connected in delta, and a 3-phase load is connected between the three junction terminals. An example using 3-phase bridge rectifiers is shown in Figure 9.1. Extending this principle, a number n of rectifying groups can be connected in a mesh or polygon, from which n-phase output can be obtained. A practical case would be four rectifiers connected in a square, giving 4-phase output (such as for driving a "2-phase" motor). However, the following analysis is particular to the 3-phase case and covers only those features that differ from the forward-reverse cycloconverter.

The most attractive application for the delta-rectifier circuit is in relatively small motor drives, where reasonable performance can be obtained with fewer thyristors than in a forward-reverse converter. The circuit of Figure 9.1 requires only 18 thyristors. A forward-reverse converter using 3-phase bridge rectifier modules requires two such modules per output phase (see Figure 1.2) or six modules (36 thyristors) for 3-phase output. There is a similar saving in the gate pulse generator circuits for the delta arrangement. The disadvantages are poorer input power factor, greater fluctuation in reactive power at a lower frequency, and the presence of circulating current around the delta connection.

After describing the basic operation in Section 9.1, an equivalent circuit is derived in Section 9.2. The effect of high frequency ratio and correction factors for use in the equivalent circuit are discussed in Section 9.3, and the procedure for computing the displacement power factor is presented in Section 9.4. The ripple voltage around the delta and methods of limiting the resulting circulating current are investigated in Sections 9.5 and 9.6.

9.1 BASIC OPERATION

The firing angles of the thyristors are modulated in the same manner as those in a single polarity group of a forward-reverse converter. If the rectifiers are regarded as positive polarity groups, then the equations governing the firing angle modulation for sinusoidal output

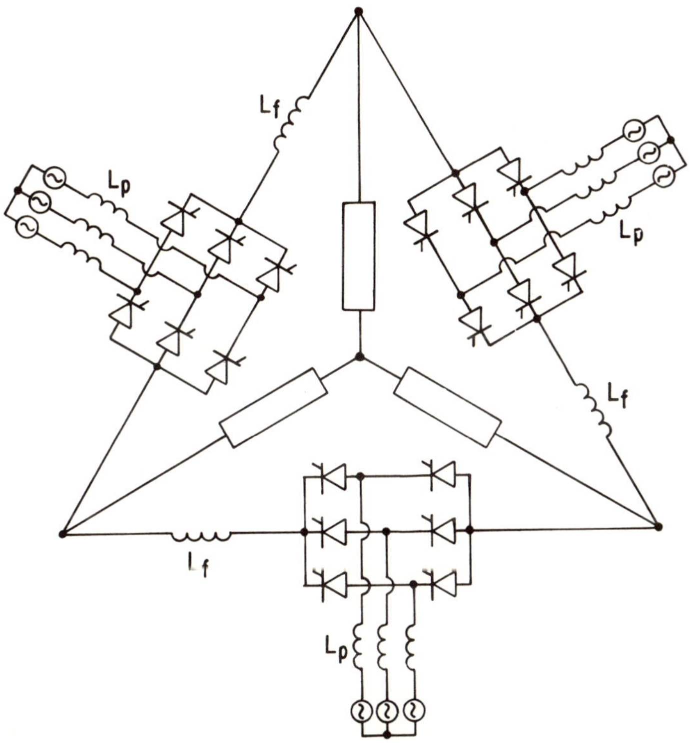

Figure 9.1. Delta-rectifier cycloconverter composed of 3-phase bridges (shown with wye-connected load).

are Equations 4.2 to 4.4. Generally, gate firing pulses are applied to the thyristors at all times, that is, there is no suppression during the intervals when the thyristors are not required to carry load current. This fact allows ripple-frequency components in the pulsating direct current that circulates around the delta. However, the elimination of suppression logic and the need to generate only one polarity of firing signals greatly simplifies the control circuitry, which is a major reason for the attractiveness of the delta-rectifier cycloconverter.

Nonsinusoidal operation that produces voltage components at the third harmonic of the output frequency is undesirable, because these components will add in phase around the delta, increasing the circulating current. However, the easily generated trapezoid, Control Method 8A in Table 6.1, contains no third harmonic and is acceptable.

The magnitude of the circulating currents is limited by the effective output impedance resulting from the rectifier source reactance and can be further reduced by adding filter reactors L_f, as shown in Figure 9.1. Also, the 3-phase modulating signals should be well balanced and should be biased to favor a negative average voltage, rather than a positive average that could cause a large direct current to circulate.

If all parasitic circulating currents are neglected, then at any instant the converter operates as an open-delta supply to the load, as represented by the equivalent circuit in Figure 9.2. The position of the "open" phase rotates every 120° of the output cycle, and the currents supplied by each rectifier are shown in Figure 9.3, together with the ac current in one phase of a wye- or delta-connected load. The parameters of the various currents are given in terms of the peak current A_m in a wye-connected load, which is $\sqrt{6}$ times the rms current A_D in a delta load.

Current delivered by each rectifying group:

dc Component,

$$I_d = \frac{3A_m}{2\pi}. \tag{9.1}$$

Fundamental component (this, of course, is the peak current in a delta connected load),

$$\begin{aligned} A_{1m} &= A_m \frac{2}{\pi} \int_0^{2\pi/3} \cos\left(x - \frac{\pi}{6}\right) \cos x \, dx \\ &= \frac{A_m}{\sqrt{3}}. \end{aligned} \tag{9.2}$$

Total rms value,

$$\begin{aligned} &= A_m \sqrt{\frac{1}{\pi} \int_0^{2\pi/3} \sin^2 x \, dx} \\ &= A_m \sqrt{\frac{1}{3} + \frac{\sqrt{3}}{8\pi}}. \end{aligned}$$

Current in each thyristor:

Peak value,

$$J_m = A_m;$$

Average value,

$$J_0 = \frac{3A_m}{2\pi p}.$$

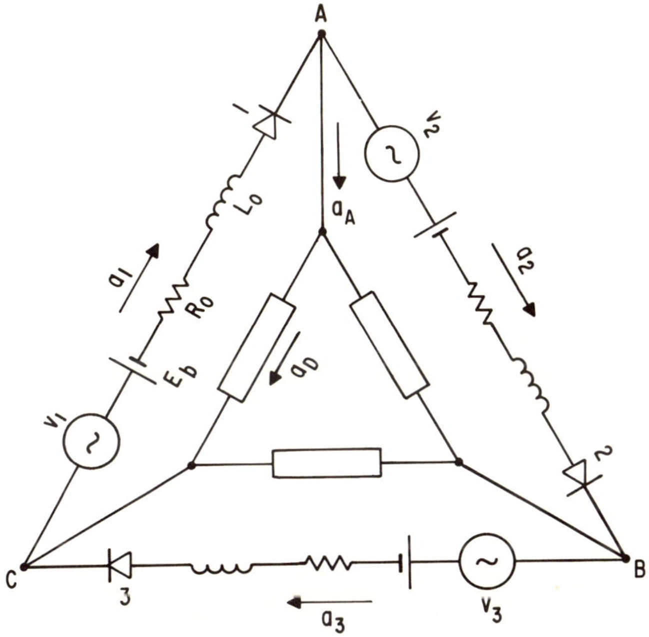

Figure 9.2. Equivalent circuit of delta-rectifier cycloconverter (shown with delta-connected load).

rms Value,

$$J_{e0} = \frac{A_m}{\sqrt{p}}\sqrt{\frac{1}{3} + \frac{\sqrt{3}}{8\pi}}$$

$$= \frac{0.6342\, A_m}{\sqrt{p}} .$$

Current in supply line to individual rectifier:

rms Value,

$$I_{e0} = J_{e0}\sqrt{n_w} .$$

9.2 FUNDAMENTAL AC EQUIVALENT CIRCUIT

The inclusion of diodes causes the equivalent circuit in Figure 9.2 to exhibit the nonlinear operation inherent in the actual delta-rectifier arrangement, particularly the current waveforms in Figure 9.3. For use in phasor diagrams and calculations, it is desirable to derive a linear equivalent circuit that is valid for the fundamental ac component of current. Consider the pseudo-instantaneous voltage between terminals B and C in Figure 9.2. When rectifier 3 is conducting, this voltage

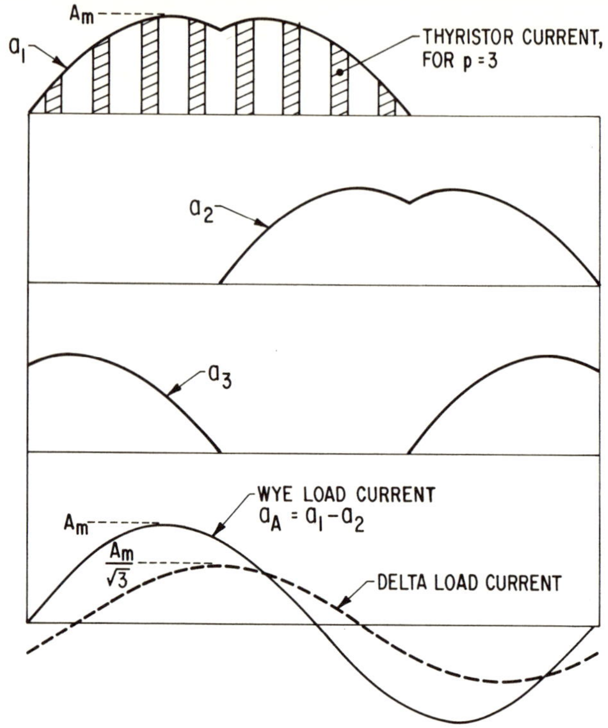

Figure 9.3. Current delivered by the three rectifiers and in wye or delta load.

is given by

$$v_{CB} = v_3 - E_b - R_0 a_3 - L_0 \frac{da_3}{dt}.$$

During the 120° in which phase 3 is idle, the terminal voltage v_{CB} is determined by the other two phases,

$$\begin{aligned} v_{CB} &= -(v_{AC} + v_{BA}) \\ &= -(v_1 + v_2) + 2E_b + R_0\,(a_1 + a_2) + L_0 \frac{d(a_1 + a_2)}{dt}. \end{aligned}$$

Since, for balanced 3-phase control signals,

$$v_3 = -(v_1 + v_2),$$

then $$v_{CB} = v_3 - \left[\frac{1}{-2}\right] E_b - R_0 \left[\frac{a_3}{-(a_1 + a_2)}\right] - L_0 \frac{d}{dt}\left[\frac{a_3}{-(a_1 + a_2)}\right],$$

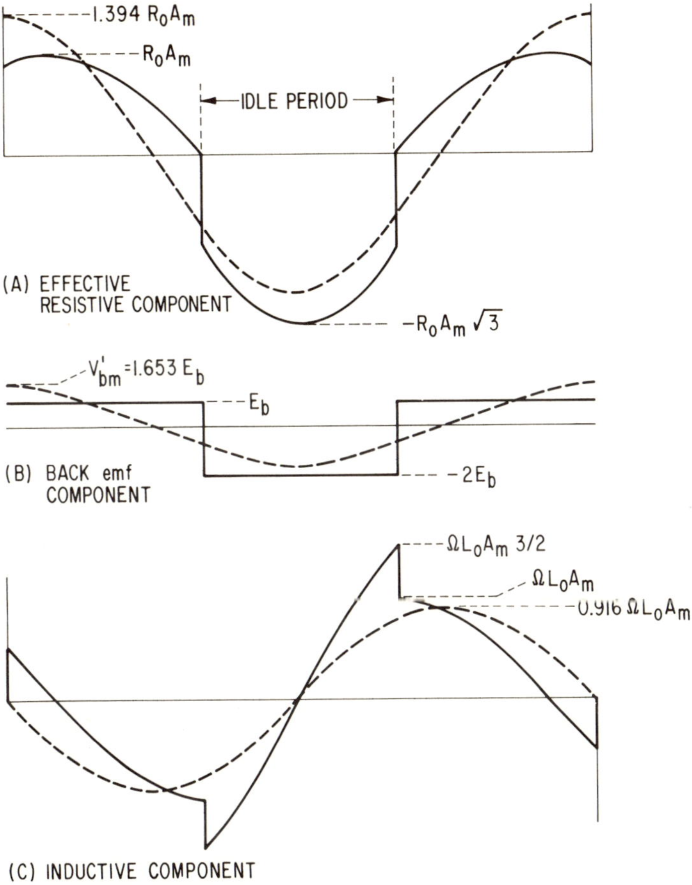

Figure 9.4. Components of voltage drop in delta-rectifier cycloconverter (phase 3).

where the upper quantities in square brackets are taken when phase 3 is conducting, and the lower quantities are taken when phase 3 is idle.

If one assumes currents as in Figure 9.3, the three components of the voltage drop are shown in Figure 9.4. The fundamental components of these waveforms are indicated by dashed lines in Figure 9.4 and are calculated as follows: The peak voltages are defined by the left-hand side of the equations in terms of the linear equivalent delta circuit parameters R_0', V_0', and L_0', which are thereby related to the parameters of the rectifier circuit, Figure 9.2. The effective resistive component is

$$R_0'A_{1m} = R_0 A_m \frac{2}{\pi}\left[\int_0^{2\pi/3} \cos\left(x - \frac{\pi}{6}\right)\cos x \, dx + \int_{2\pi/3}^{\pi} \sqrt{3}\cos x \cos x \, dx\right]$$

$$= R_0 A_m\left(\frac{2}{\sqrt{3}} + \frac{3}{4\pi}\right)$$

$$= 1.394\, R_0 A_m. \tag{9.3}$$

In view of Equations 9.3 and 9.2,

$$R_0' = \left(2 + \frac{3\sqrt{3}}{4\pi}\right)R_0 = 2.413\, R_0\,. \tag{9.4}$$

The back emf component is

$$V_{bm}' = E_b \frac{2}{\pi}\left[\int_0^{2\pi/3} (1)\cos x \, dx + \int_{2\pi/3}^{\pi} (-2)\cos x \, dx\right]$$

$$= E_b \frac{3\sqrt{3}}{\pi} = 1.653 E_b$$

and the rms value is

$$V_b' = \frac{V_{bm}'}{\sqrt{2}} = 1.17 E_b = 1.3 V_b\,. \tag{9.5}$$

The inductive component is

$$\Omega L_0' A_{1m} = \Omega L_0 A_m \frac{2}{\pi}\left[\int_0^{2\pi/3} \sin\left(x - \frac{\pi}{6}\right)\sin x \, dx + \int_{2\pi/3}^{\pi} \sqrt{3}\sin x \sin x \, dx\right]$$

$$= \Omega L_0 A_m\left(\frac{2}{\sqrt{3}} - \frac{3}{4\pi}\right)$$

$$= 0.916\Omega L_0 A_m\,.$$

Therefore

$$L_0' = \left(2 - \frac{3\sqrt{3}}{4\pi}\right)L_0 = 1.587 L_0\,. \tag{9.6}$$

Thus the fundamental frequency output characteristics of the converter can be calculated by means of a linear equivalent delta circuit, one phase of which is shown in Figure 9.5, and its phasor diagram in Figure 9.6. If a wye equivalent circuit is desired, the voltages must be divided by $\sqrt{3}$, the currents multiplied by $\sqrt{3}$, and the impedances divided by 3. To account for the frequency ratio correction factors

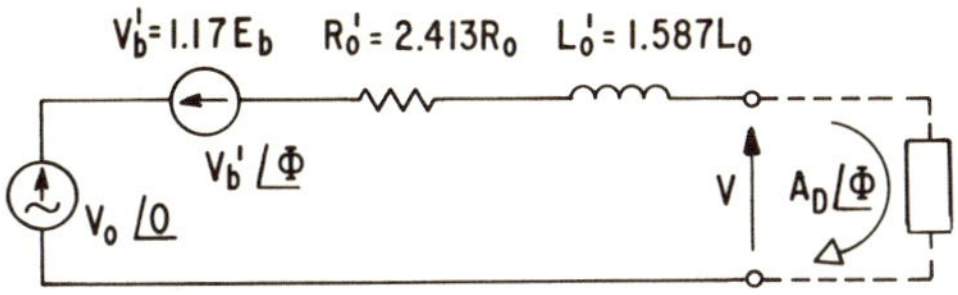

Figure 9.5. One phase of linear equivalent delta circuit.

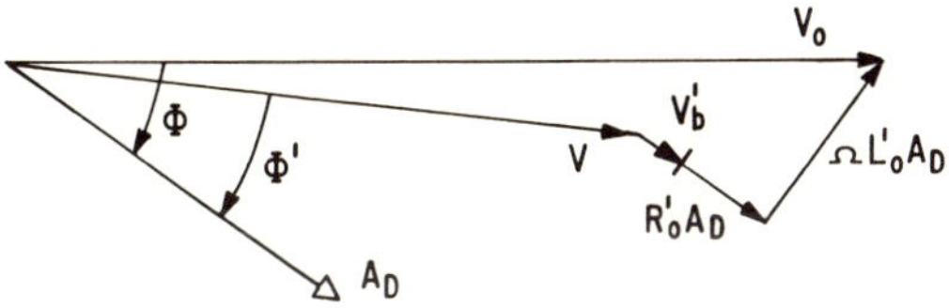

Figure 9.6. Phasor diagram for linear equivalent circuit.

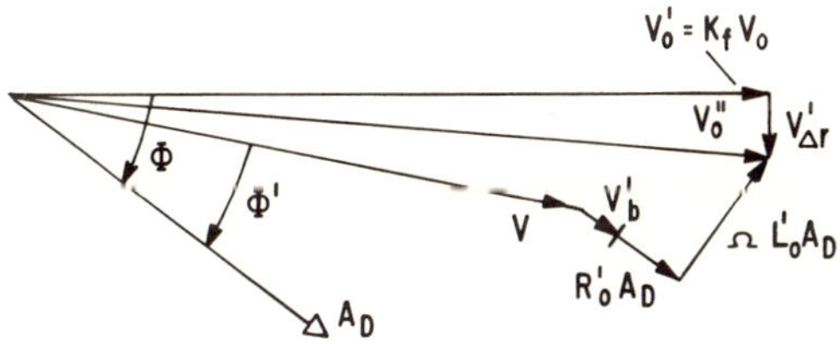

Figure 9.7. Phasor diagram including frequency ratio correction factors.

derived in the next section, the phasor diagram, Figure 9.6, can be modified as shown in Figure 9.7.

9.3 CORRECTION FACTORS FOR HIGH FREQUENCY RATIO

Since the delta-rectifier cycloconverter consists of positive current groups only, the no-load output voltage according to the theory developed in Chapter 5 is the v'_{0p} component of Figure 5.3 (A). All three phase voltages are drawn in Figure 9.8, and the fundamental components $v'_0 = K_f v_0$ are shown by dashed lines. The second harmonic distortion components $K_\Delta v_0$ of the three voltages combine to produce a sixth harmonic of the output frequency, shown at the bottom of Figure 9.8, which probably contributes to the circulating current around the delta. Thus, the reduction factor K_f remains effective in the delta-rectifier converter but not the current-dependent voltage components $V_{\Delta a}$ and $V_{\Delta b}$ that are produced in the forward-reverse converter.

The component $V_{\Delta r}$ due to variation of the equivalent resistance should still be determined by the pseudo-instantaneous Equation 5.29, which contains a term proportional to the current and a term due to saturating reactors, if such are used. Based upon the same consideration of conducting and idle periods for each rectifier, which was described in Section 9.1, the contributions of these two terms are shown in Figure 9.9 for purely resistive load and in Figure 9.10 for purely inductive load. As in the case of the forward-reverse converter, the fundamental components of both terms lead the no-load voltage by 90°, for all types of load. The fundamental rms value of the voltage $V'_{\Delta r}$ can be expressed in terms of factors K_a and K_s, for use in the delta equivalent circuit,

$$V'_{\Delta r} = kspF(K_a L_c A_D + K_s L_u I_s). \tag{9.7}$$

By Fourier analysis of the waveforms in Figures 9.9 and 9.10, the factors are

	K_a	K_s
Resistive/regenerative load	$\frac{9}{4\pi} = 0.716$	$\frac{3}{\sqrt{2}\pi} = 0.675$
Inductive/capacitive load	$1 + \frac{9\sqrt{3}}{4\pi} = 2.24$	$\frac{6-\sqrt{3}}{\sqrt{2}\pi} = 0.961$
Mean value	1.48	0.82

To simplify computation, the mean values of the factors can be used with any type of load, because their effect is generally small. The phasor diagram including the frequency ratio correction factors is shown in Figure 9.7 and can be solved by a set of equations similar to Equations 4.25 to 4.30.

9.4 DISPLACEMENT POWER FACTOR

As mentioned in Section 6.3, the average displacement power factor of the delta-rectifier cycloconverter can be calculated from the ratio P_X/P_R given by Equation 6.6 with an expression for the nonsinusoidal current delivered by a rectifier in place of $\sin(x - \Phi)$. As indicated in Figure 9.11, it is convenient to select the x or time origin at the peak of the output voltage wave, which becomes in normalized form

$$\bar{v}_0 = k \cos x.$$

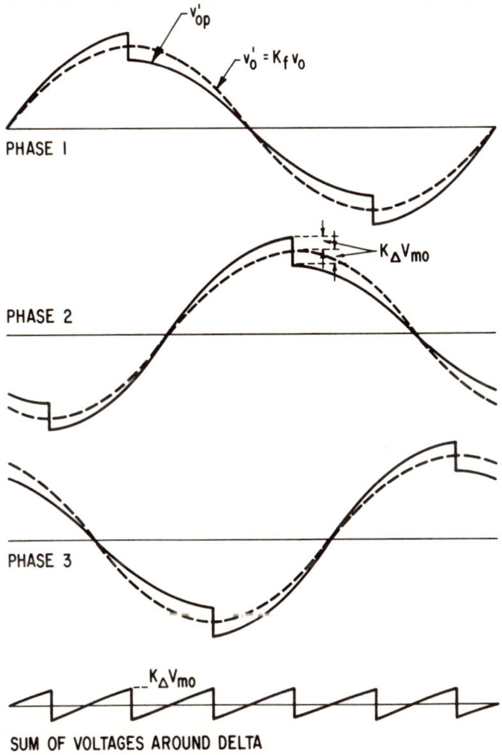

Figure 9.8. Rectifier output voltages including frequency ratio distortion.

The current waveform can then be expressed as

$$a = A_m \cos\left(|x - \Phi| - \frac{\pi}{6}\right)$$

over the range

$$\Phi - \frac{2\pi}{3} < x < \Phi + \frac{2\pi}{3}.$$

Equation 6.6 therefore becomes

$$\frac{P_X}{P_R} = \frac{\int_{\Phi-2\pi/3}^{\Phi+2\pi/3} \cos\left(|x-\Phi| - \frac{\pi}{6}\right)\sqrt{1 - k^2\cos^2 x}\; dx}{\int_{\Phi-2\pi/3}^{\Phi+2\pi/3} \cos\left(|x-\Phi| - \frac{\pi}{6}\right) k\cos x\; dx}.$$

The value of the integral in the denominator is $(\pi/\sqrt{3})\, k \cos \Phi$, but

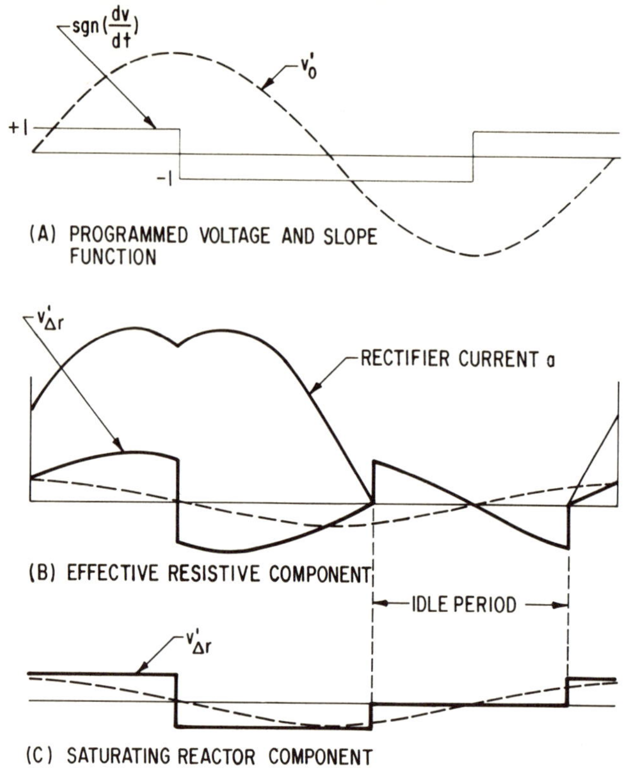

Figure 9.9. Components of distortion voltage $v_{\Delta r}'$ with resistive load.

the numerator is best integrated numerically when $k < 1$. For accuracy, the interval of integration should be divided into regions bounded by discontinuities in the slope of the function, at $x = \Phi$ because of the current and (when $k = 1$) at $x = 0$ and π because of the term $\sqrt{1 - \cos^2 x}$. However, if the interval is not divided so that the numerical method approximates the function by a smooth curve, the error is found to be quite small. The computed displacement power factor is plotted in Figure 9.12.

The maximum power factor of the delta-rectifier cycloconverter is seen to be 0.688, compared with 0.844 for a forward-reverse converter with the same sinusoidal output voltage. Note that each rectifier must operate in the inverting mode for at least 60° of each output cycle, because the current flows for 240° while the voltage can have

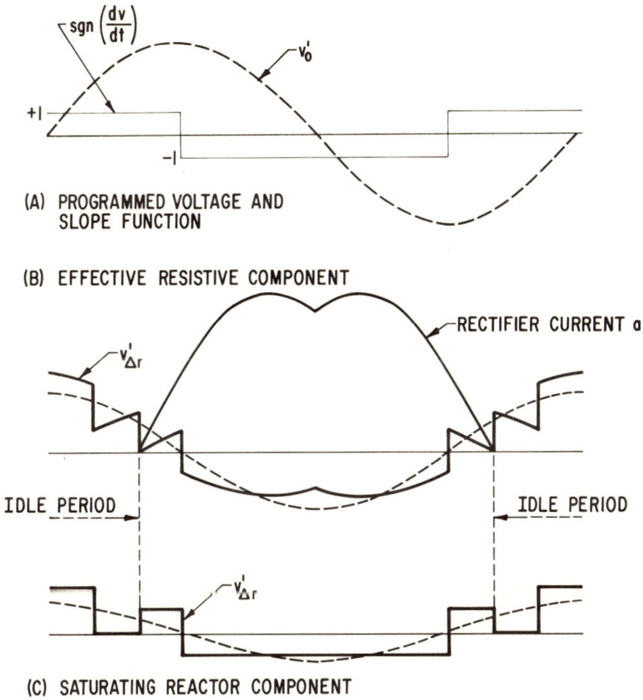

Figure 9.10. Components of distortion voltage $v'_{\Delta r}$ with inductive load.

the same polarity for only 180°. That is, the real power P_r in Figure 9.11 must be negative for at least 60°, a fact that explains the poor power factor of the delta circuit.

Also, the reactive power P_x fluctuates at the output frequency instead of its second harmonic (compare Figures 9.11 and 4.2), or the total reactive power drawn from the supply system for balanced 3-phase output fluctuates at the third harmonic instead of the sixth harmonic. The magnitude of fluctuation is also greater than in the case of the forward-reverse converter. For example, with low output voltage k approaches zero and the firing angle of all thyristors approaches 90°, therefore the magnitude of current in each input phase will be the direct addition of the contributions from each output phase. Thus, the total reactive power will have the same shape as the sum $a_1 + a_2 + a_3$ of the currents in Figure 9.3, which has the waveform of a 3-pulse diode rectifier, giving a min/max ratio of 1/2.

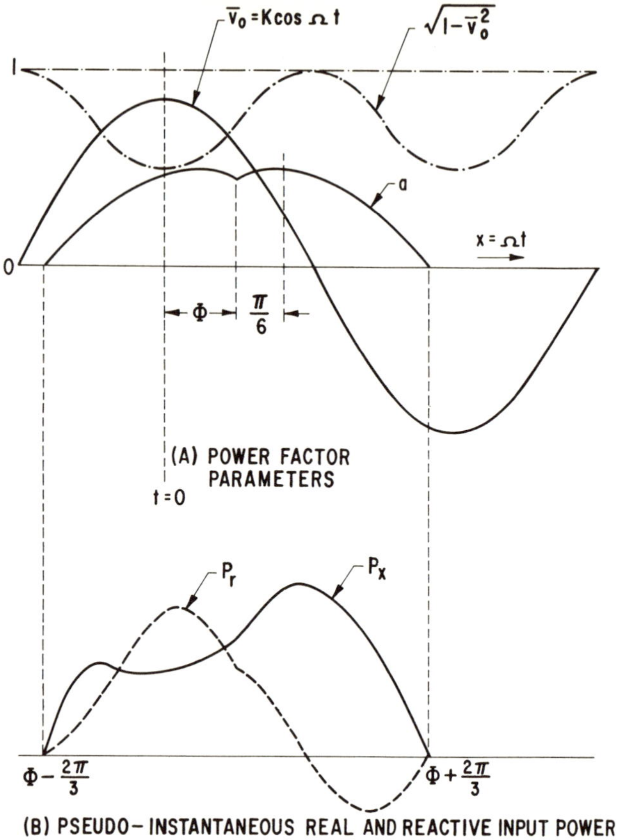

Figure 9.11. Input power of delta-rectifier cycloconverter with sinusoidal output.

9.5 LIMITING THE CIRCULATING CURRENT

The component of current circulating around the delta due to the output voltage ripple (neglecting the frequency ratio effect) is at the qth harmonic of the supply frequency. If the individual rectifier supplies have the same set of phases, the maximum ripple occurs when the output voltage is zero, or $k = 0$, for which $\alpha = \pi/2$ at all times. The integral of the ripple voltage produced by each rectifier is then given by Equation 2.18, and the total sum around the delta is three times that value. If it is desired to limit the peak-to-peak current caused by this voltage to a value A_Δ, which may be specified as a fraction of the peak load current A_m, then the size of the filter inductance can be

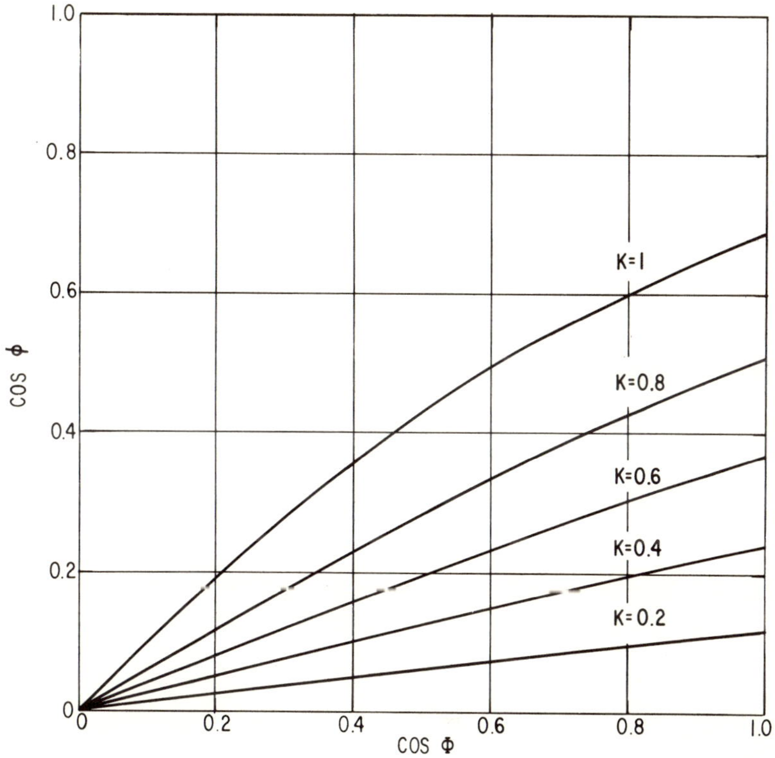

Figure 9.12. Variation of input power factor cos ϕ with output power factor cos Φ and modulation fraction k. Delta-rectifier cycloconverter with sinusoidal output voltage.

obtained by inverting Equation 2.19. Since the total output inductance L_0, which includes the source, is effective in suppressing the ripple current, it is required that

$$L_0 = L_f + sL_c \geqslant \frac{E'_m\,[1 - \cos(\pi/q)]}{\omega A_\Delta}.$$

9.6 PRINCIPAL COMPONENT OF RIPPLE VOLTAGE AROUND DELTA

To determine the manner in which the circulating ripple current varies with the modulation factor k, a computer program was adapted from the previously discussed program to calculate the principal (qth) harmonic component of the ripple voltage summed around the delta. Although this component does not relate directly to the actual pulsating dc circulating current, it serves as a comparative factor in guiding the circuit designer.

Considering the converter to consist of all positive current groups, the firing angles of the three phases at a time phase $x_i = \Omega t_i$ are given by

$$\alpha_{i1} = \text{Cos}^{-1}[\bar{v}_0^{(x_i)}]; \tag{9.8}$$

$$\alpha_{i2} = \text{Cos}^{-1}[\bar{v}_0^{(x_i+2\pi/3)}]; \tag{9.9}$$

$$\alpha_{i3} = \text{Cos}^{-1}[\bar{v}_0^{(x_i-2\pi/3)}]. \tag{9.10}$$

Use of the principal value of the inverse cosine function (denoted by capitalization) ensures that when $\bar{v}_0$ is positive, the corresponding value of α is in the range 0 to $\pi/2$ radians. Negative values of $\bar{v}_0$ yield values of α in the range $\pi/2$ to π radians. Substituting these firing angles in Equations 7.1 and 7.2, one obtains corresponding values of the principal harmonic coefficients. The following integral is then evaluated numerically, to yield one-third of the resultant rms principal harmonic around the delta:

$$C_{1e} = \frac{1}{3}\sqrt{\frac{3}{\pi}\int_0^{\pi/3}[(A_{11} + A_{12} + A_{13})^2 + (B_{11} + B_{12} + B_{13})^2]\,dx}.$$

Although an integration interval of $\pi/6$ is sufficient in this case, an interval of $\pi/3$, which is required for the integral described later, is used so that the two can be computed during the same run.

The use of symmetrically place-displaced sets of supplies for each of the three converters has been proposed to reduce the circulating current. To estimate the benefit obtained from such a more complex arrangement, the computer program was written to include calculation of the resultant principal harmonic in this case also. From Equation 2.20, the mqth harmonic voltage produced by a converter fed from a reference set of supply phases can be written as

$$e_m(\omega t) = a_m \sin(mq\omega t) + b_m \cos(mq\omega t).$$

Then, the mqth harmonic voltage produced by another converter fed from a second set of supply phases, which are displaced to lead the reference set by an angle δ at the supply frequency, may be written with respect to the same time origin, which is a point of natural commutation of the first set,

$$\begin{aligned} e_m(\omega t + \delta) &= a_m \sin mq(\omega t + \delta) + b_m \cos mq(\omega t + \delta) \\ &= (a_m \cos mq\delta - b_m \sin mq\delta) \sin (mq\,\omega t) \\ &\quad + (a_m \sin mq\delta + b_m \cos mq\delta) \cos (mq\,\omega t). \end{aligned}$$

Thus, for $m = 1$, the principal harmonic coefficients produced by the second converter referred to the same quadrature phases ($\sin q\omega t$ and $\cos q\omega t$) as the first converter are

$$\begin{aligned} A_1(q\delta) &= A_1 \cos q\delta - B_1 \sin q\delta; \\ B_1(q\delta) &= A_1 \sin q\delta + B_1 \cos q\delta. \end{aligned}$$

Now, the desired symmetrical phase shift to produce some harmonic cancellation is

$$\delta = \pm \frac{2\pi}{3q}$$

or

$$q\delta = \pm \frac{2\pi}{3},$$

where the + sign may be taken for the second converter and the − sign for the third converter (or vice versa; the relative displacement of the supplies and the output phase rotation does not matter). Since the firing angles of the three converters are still given by Equations 9.8 to 9.10, the integral to be evaluated is

$$C'_{1e} = \frac{1}{3}\sqrt{\frac{3}{\pi}\int_0^{\pi/3}\left[\left\{A_{11}(0) + A_{12}\left(\frac{2\pi}{3}\right) + A_{13}\left(-\frac{2\pi}{3}\right)\right\}^2 + \left\{B_{11}(0) + B_{12}\left(\frac{2\pi}{3}\right) + B_{13}\left(-\frac{2\pi}{3}\right)\right\}^2\right] dx}.$$

The computed values of C_{1e} and C'_{1e} as functions of k and q are plotted in Figure 9.13 for $q = 2$ and 6. For supplies having the same phase, the ripple is a maximum for $k = 0$, and about 50% of that maximum for $k = 1$, with a minimum at some intermediate value of k (for $q > 2$). The major effect of using displaced-phase supplies is to eliminate the net qth harmonic around the delta for $k = 0$, although ripple corresponding to a pulse number of $3q$ remains. However, as the modulation factor k is raised, the qth harmonic reappears and soon increases to a level comparable to that obtained with supplies from the same set of phases. Thus, except for $q = 2$ where a 3-phase

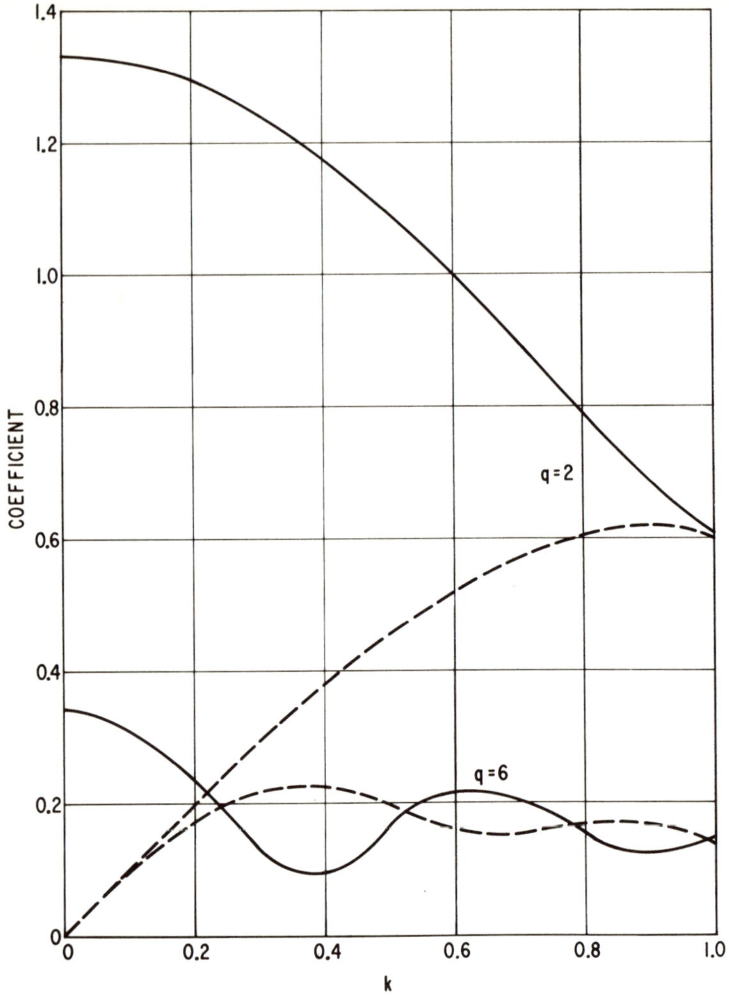

Figure 9.13. Coefficient of principal harmonic voltage in delta-rectifier cycloconverter (one-third of sum around delta) as a function of the modulation fraction k and pulse number q. Solid lines : supplies having same phase. Dashed lines: symmetrically displaced supplies.

system naturally gives the required phase shift and the improvement is considerable, the value of supply phase displacement is doubtful. However, it may be convenient and contribute to separation of the source reactance of the three converters if the source is an isolated generator.

10 Computer-Aided Cycloconverter Analysis and Design

The analysis of cycloconverters has been carried out in terms of assumed values of supply voltage, load current, inductances, and resistances. The results of importance in performance calculation can be incorporated in a computer program to aid the designer. Essentially, selected equations from the previous sections are strung together in logical order for step-by-step computation and iterated where necessary. An outline flow chart is shown in Figure 10.1. This program solves a detailed equivalent circuit, including correction factors for high frequency ratio, and makes other performance calculations. The supply voltage, load current, power factor of the load and both the input and output frequencies, as well as the circuit configuration and all circuit impedances, are given parameters. Any components that are omitted (such as a filter L_f) or are neglected for preliminary studies (such as the resistance R_c) should be set equal to zero.

The equivalent circuit phasor relationships are solved for the output voltage, iterating to account accurately for the effect of internal phase angle on the frequency ratio correction factors, for a given modulation fraction. If the design option discussed in Section 10.1 is selected, the modulation fraction or supply voltage is changed, and the phasor calculations are iterated until the computed output voltage is within a specified tolerance of the desired value.

Converter performance calculations are then made, particularly the input/output power relations and losses. In the subroutine for calculating the average reactive input power by numerical integration, another subroutine is invoked. This subroutine calculates the pseudo-instantaneous firing angle for a given normalized voltage signal and then the overlap angle appropriate to the pseudo-instantaneous current, from which the factor p_x giving the reactive component of input power is obtained. It also affords an opportunity to check whether there is a commutation failure or whether operation is not regular (mode 1) at any of the instants selected by the numerical integration routine.

The equivalent circuit approach has been found to yield satisfactory results for the design of cycloconverters used in variable-speed,

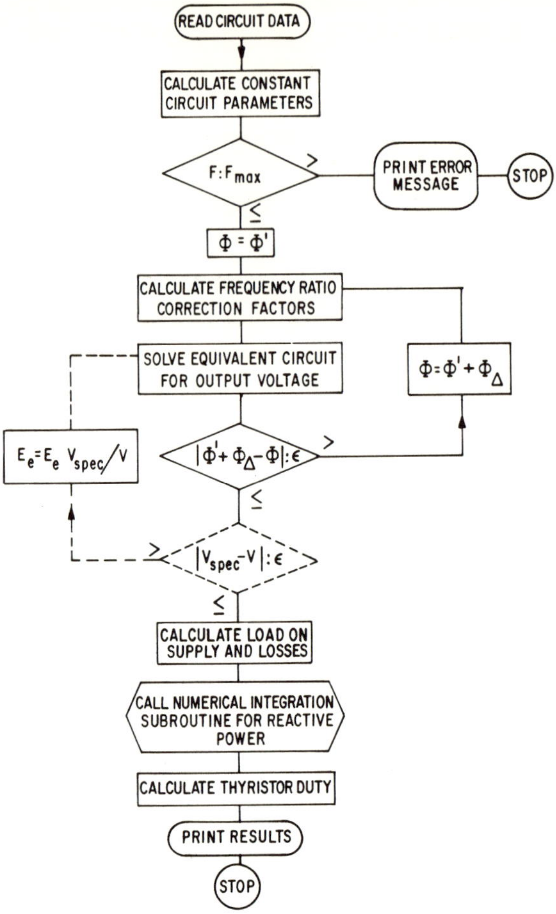

Figure 10.1. Flow chart for computer analysis of cycloconverter performance.

constant-frequency (VSCF) systems for aircraft 400 Hz power supplies. In this application, extensive filtering is necessary to obtain the high-quality waveform demanded by the system specifications. Therefore, the converter current waveform probably approaches the sinusoidal shape assumed in the standard analysis.

However, in motor drive applications, filter reactors are usually omitted in the interest of economy. The filtering action provided by the inductance of the motor windings is generally quite small. Since the major portion of a motor load appears as a counter emf to the cycloconverter, the current waveform in many instances has a higher ripple content than the voltage and can become discontinuous over a considerable portion of the output cycle. Thus, the basic analytical assumption that the current is sinusoidal and continuous is grossly violated. It appears that unsuppressed ripple currents result in high losses occurring in the motor, converter, and supply transformer, as well as an increase in reactive power. The present analysis is inadequate to account for such losses.

It is believed that the capability of an equivalent circuit approach to the evaluation of cycloconverters has been exhausted by the present study. If more accurate theoretical results are desired, a direct simulation of the converter circuit and its load should be made, using an analog,[34] digital, or hybrid[55] computer.

10.1 SYNTHESIS BY ITERATED PERFORMANCE COMPUTATION

The computer program can, of course, be modified by the user to better suit the objectives of a particular design study. In the example shown by dashed lines in Figure 10.1, the supply voltage is adjusted by iteration to obtain a specified output voltage. This option is useful in applications such as VSCF systems and vehicle traction motor drives, where the voltage of an isolated generator is regulated to obtain the desired output. It also could be used to specify the secondary voltage of a transformer in other applications, if the other given parameters represent the worst case with respect to voltage. Another option would perform a similar iteration of the signal voltage (modulation fraction) to obtain a specified output voltage, as required in designing converters to operate from a fixed-voltage supply, such as the 60 Hz utility.

A thyristor-limited design can be obtained by suitable program modification. Equations in the analysis give various critical levels of electrical stress to which the thyristors are subjected, such as peak voltage, rms, average and surge currents, and turn-on di/dt. In a thyristor-limited design of the circuit, the supply voltage, load current, and inductances are not assumed but are chosen so that the actual thyristor stresses match the maximum values permitted by the thyristor specifications, reduced by any desired safety factors.

Although the results will not generally match any actual source or load (i.e., in a practical circuit, some thyristor stresses will not be as high as their limit values), this procedure serves to determine the maximum performance of a given circuit configuration using the specified thyristors. This is useful in comparing alternative thyristors that may be considered for cycloconverter duty in terms of output kVA per thyristor or some other criterion. The process also emphasizes those parameters that the designer needs to know and their relative importance in determining converter performance.

As an illustration, suppose it is desired to compare different cycloconverter circuit configurations with respect to the maximum output that can be obtained and how the load voltage will vary with the output frequency, when each circuit uses the same type of thyristor. For an assumed constant load current at the maximum value determined by the thyristor rating and at some prescribed power factor, the performance of each converter is computed at selected frequencies, for the maximum possible output voltage. Such computer calculations were made for 3-phase single-way converters connected in wye with third harmonic modulation of the neutral point voltage (Control Method 7 in Table 6.1), for 3-phase bridge converters with pure sinusoidal modulation, and for a delta-rectifier circuit composed of 3-phase bridges.

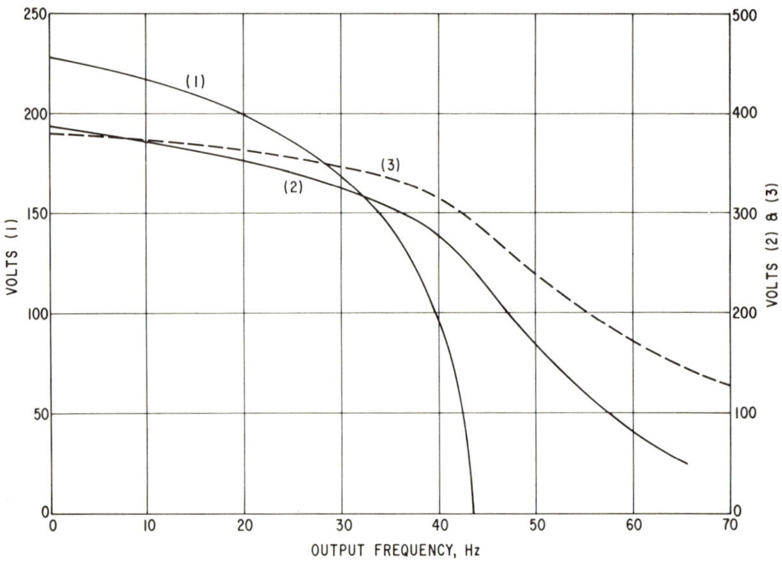

Figure 10.2. Computed variation of load voltage with output frequency (example).
(1) 3-phase single-way converters connected in wye, with optimum nonsinusoidal modulation.
(2) 3-phase bridge converters with sinusoidal modulation.
(3) 3-phase bridge delta-rectifier circuit with sinusoidal modulation.

The load voltages in each case are plotted in Figure 10.2 as a function of frequency. Since the bridge converters consist of two commutating groups in series, the three curves are directly comparable in relative performance when the scale for the single-way circuit is double that of the bridge circuits, as is drawn. The higher relative output of the single-way circuit before its frequency limit is reached is a virtue of the nonsinusoidal modulation. The bridge circuits have a higher frequency limit. The higher voltage of the delta circuit at higher frequencies is attributed to the elimination of the distortion voltage terms.

For 3-phase output, a wye connection of 3-phase single-way forward-reverse converters appears to be an optimum configuration when the frequency ratio is 1/3 or less. If the voltage of the load is matched with respect to the supply, isolation transformers or 6-terminal load windings can be avoided. Furthermore, an improvement can be obtained by employing some form of flat-topped voltage modulation, because the third harmonic is eliminated by the wye connection. This results in a higher fundamental component of output voltage, improved input power factor, and less ripple in the output.

10.2 THYRISTOR SPECIFICATION

The thyristors employed in cycloconverters are subjected to a peculiar current waveform, which makes interpretation of the current rating rather difficult. Thyristor manufacturers have developed computer programs to calculate current rating with unusual duty cycles. A thyristor that has a high surge current capability relative to its steady-state capacity is desirable.

In cycloconverters and other phase-controlled converters, the commutating inductance determines the voltage regulation of the circuit, the *di/dt* of the thyristors, and the possible surge current. Generally, it is found that the surge current rating of the thyristors is the limiting factor in determining the value of commutating inductance required for safe operation. Typically, a relatively modest *di/dt* capability is sufficient to achieve a tolerable amount of regulation. However, a large value of commutating inductance results in poor voltage regulation and poor power factor, as is well known in the rectifier art. The poor duty cycle in a cycloconverter accentuates the problem.

The after-commutation *dv/dt* is the slope of the line-to-line voltage waveform, which is quite modest even at high supply frequencies if spikes produced by other commutations are not a problem. However, in cycloconverters having thyristors connected in inverse-parallel, the reverse *dv/dt* on one thyristor following its turnoff appears as forward *dv/dt* on its mate, which has been nonconducting during the previous interval. This blocking *dv/dt* is much higher than the after-commutation *dv/dt* and must be limited by RC snubber circuits. The *dv/dt* rating of the thyristors determines the size of the RC snubber circuits after the inductances have been determined. With high supply frequencies, the snubber losses can become excessive, and it may be desirable to use saturating reactors to increase the effective inductance when the current is near zero.

The turn-off time of the thyristors does not directly limit any circuit components but affects the permissible range of operating conditions, particularly the practical maximum supply frequency. It also limits the terminal voltage for regenerative loads, but many applications do not require regeneration at the full power capacity of the circuit. Another effect of turn-off time is to require a delay in the changeover of the current polarity blanking logic in those circuits where this method of control is used.

If, in a forward-reverse cycloconverter, it is assumed that circuit malfunctions, such as misfiring, are liable to result in repetitive intergroup faults, then the intergroup reactance should be sized to limit the peak current to some repetitive surge rating of the thyristors. As for ordinary rectifiers, optimum thyristor utilization is achieved when the "commutating number" $p = 3$. Other circuit arrangement factors have a greater effect on overall circuit performance than on the thyristors in particular.

11 Applications of Cycloconverters

Cycloconverters have been used or proposed for use in most applications where power frequency conversion is desired or where such conversion offers an alternative method of achieving system or process objectives that are now usually obtained in some other way. The actual introduction of cycloconverter equipment to practical usage is governed by the development of favorable technical-economic factors for the overall system, as for any other apparatus, of course. After summarizing the various types of cycloconverter application, details of a few specific present installations are presented. A comparison with competitive frequency or speed-changing systems is made.

From the aspect of converter operation, the applications of cycloconverters can be categorized according to whether the input or output frequency is variable or constant. Usually, but not always, the input or output voltage is also variable or almost constant, according to the frequency category. Four combinations are possible, listed here with examples of utilization in systems.

1. Constant frequency input, constant frequency output: for example, links between power systems of different frequencies.

In such systems, both the input and output voltages are also constant. One of the earliest applications of cycloconverters was to link the 50 Hz 3-phase utility power system to a $16\frac{2}{3}$ Hz single-phase distribution system for electric railroads in Germany during the 1930s. However, these installations have been superseded.

2. Variable frequency input, constant frequency output: for example, VSCF generating systems for aircraft, space vehicles, and mobile power supplies.

Here, the input voltage from the generator is usually regulated to an approximately constant value so that constant output voltage can be obtained with full modulation of the converter. Details of such a system are presented in Section 11.1.

3. Constant frequency input, adjustable frequency output: for example, ac motor drives of all types.

The input voltage is also constant (60 Hz or other utility), while the output voltage is varied in approximate proportion to the output frequency. The motor may be of the synchronous or induction type, and the drive may be for a single large motor or multiple smaller motors. Besides, industrial drives, exemplified in Sections 11.2 and 11.3, cycloconverter drives have been proposed for rapid transit cars, catenary-supplied electric locomotives, and position servos, such as for a large, steerable antenna.

4. Variable frequency input, adjustable frequency output: for example, electric vehicle propulsion drives with on-board prime mover.

As in other motor drives, the output voltage is controlled as a function of the frequency. The frequency of the input generator can be allowed to vary in order to optimize the performance of the engine, for which the most efficient speed is a function of the power demand. The generator voltage may be controlled to some extent, so that the cycloconverter can be operated with a high degree of modulation where its power factor is better. Cycloconverter propulsion drives have been proposed for diesel or turbine locomotives and large off-highway wheeled or tracked vehicles.

11.1 VSCF GENERATING SYSTEM

Variable speed, constant frequency (VSCF) refers to an electrical power generating system in which the generator is driven by the prime mover at a variable speed; therefore, it produces a variable frequency output. An electric power converter is then used to produce a constant frequency for feeding to the system loads. The scheme is particularly attractive for use in aircraft, where the standard 400 Hz power is, at present, usually provided directly by a generator with a constant-speed hydraulic-mechanical drive from the aircraft engine. A VSCF system using a cycloconverter has certain advantages over the constant-speed drive system and other VSCF systems that use a dc link (rectifier-fed inverter). There is also interest in VSCF systems for mobile military power supplies, where the weight, regulation, and gear train drive of a low-frequency generator cause problems.

The following description pertains to a particular 60 kVA VSCF system developed by the General Electric Company for use in aircraft, although it is probably typical. A block diagram of the system is shown in Figure 11.1, where only one of the three output phases is drawn in detail. The generator operates over a speed range of 10,000 to 20,000 r/min, providing a corresponding variable frequency output from 1200 to 2400 Hz. Cycloconverters change this to precise 400 Hz

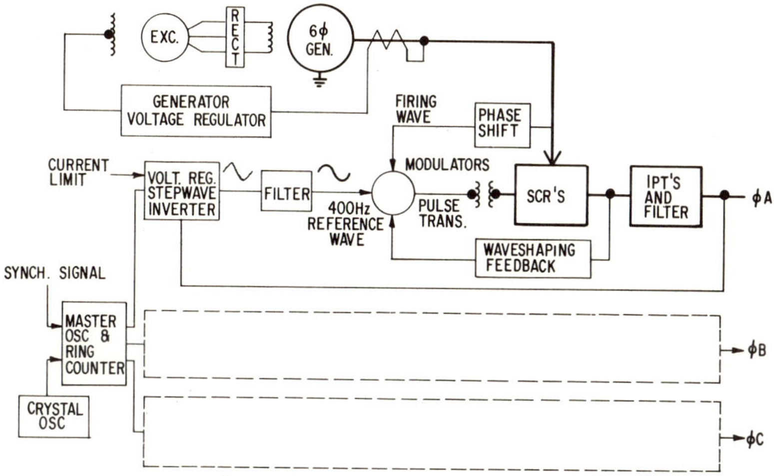

Figure 11.1. Block diagram of a VSCF generating system.

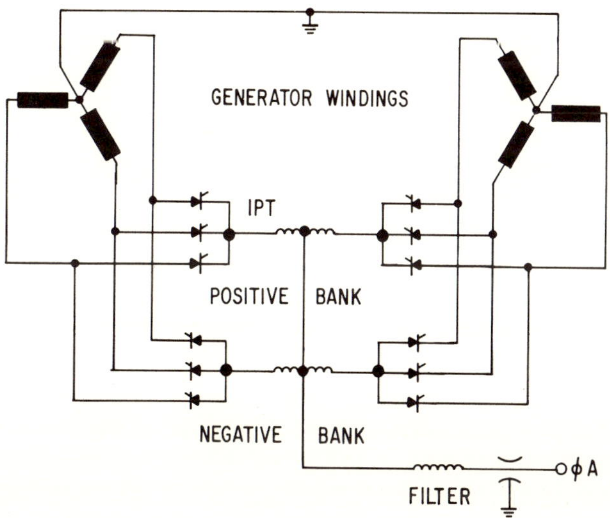

Figure 11.2. VSCF cycloconverter power circuit.

power at 115/200 V, 3-phase. A total of 36 thyristors or silicon-controlled rectifiers (SCRs in Figure 11.1) are used in the converters.

Details of the heavy-lined blocks in Figure 11.1 are presented in Figure 11.2. The generator has 6-phase windings, which feed a 6-pulse cycloconverter. Each current polarity bank of six thyristors consists of two 3-pulse commutating groups with an interphase transformer (IPT). This arrangement has the better thyristor and winding utilization of a 3-phase circuit with the ripple properties of a 6-phase circuit. The star point of the generator windings is the neutral terminal of the 400 Hz output, which is required by the system specifications. The ripple frequencies are suppressed by a simple low-pass filter consisting of a series inductance and a feed-through capacitor.

The highest output/input frequency ratio is 400 Hz/1200 Hz = 0.33, which is just within the "good" waveform region of Figure 5.12. The unfiltered waveform is similar to that drawn in Figures 1.3 and 1.4. However, low harmonics of the output frequency are introduced by nonlinearities in the converter. These are reduced by a feedback circuit, designated "waveshaping feedback" in Figure 11.1. The production of gate firing pulses for each thyristor is controlled by a microcircuit modulator in which three signals are mixed: a firing wave at the generator frequency, the 400 Hz reference signal, and the waveshaping feedback signal.

The 400 Hz reference wave is generated as a 24-step square wave using a master oscillator controlled by a very accurate crystal oscillator. The height of each step in this quasi-square wave is such that the lowest harmonic present is the 23rd. The harmonics are then removed with a relatively small active filter.

The output voltage of each phase is regulated independently by comparing it with a voltage reference and using the error signal to vary the amplitude of the reference wave. If the output voltage is high, the error signal causes the amplitude of the reference wave to be reduced, which then causes the output voltage to be reduced.

As indicated on the block diagram in Figure 11.1, a separate voltage regulator is used to hold the generator voltage essentially constant. In addition, a current limit circuit acts to hold the output current to less than approximately 3.0 per unit during faults.

The converter with its filter, plus the system control and protection circuits, are enclosed in a single compact package, Figure 11.3, weighing 58 lb or about 1 lb/kVA. Cooling air is directed across the components by a pressure plenum at the back, and exits out the front. The development of this converter took place over a number of years, and some of the work is reported in References 13 and 21.

11.2 MULTIPLE AC MOTOR DRIVE POWER SUPPLIES

One of the earliest applications of thyristor cycloconverters was to

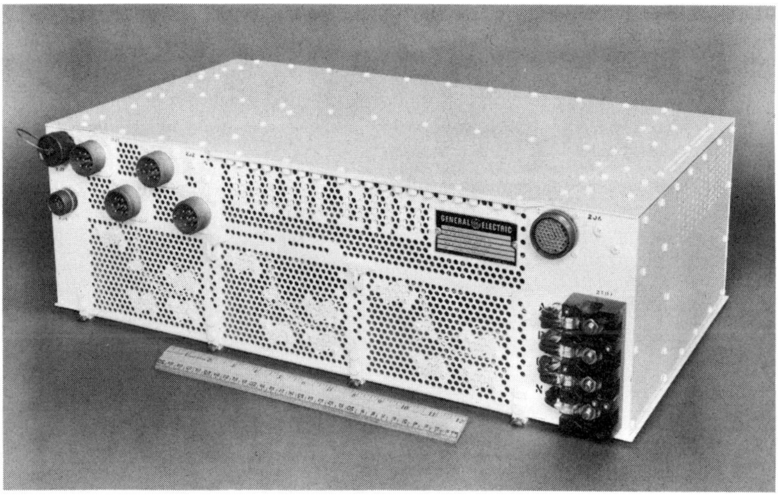

Figure 11.3. 60 kVA VSCF converter/control assembly.

power the reversing run-out table motors for a hot strip mill. A typical installation will be described; more details are given in Reference 36. Many factors were considered in selecting an ac rather than a dc drive system. Since about 300 motors of 2.6 hp each were involved, elimination of brush maintenance favored the ac drive system. The speed range was suitable for the application of cycloconverter supplies.

The induction motors were specifically designed to match the most economic converter design. The number of poles (6) was selected to require a frequency of 13 Hz at maximum speed, allowing a simple 3-pulse converter to be used. The frequency ratio 13 Hz/60 Hz = 0.22 is just into the "fair" waveform region defined in Figure 5.10, while the normal top frequency of 11 Hz is on the border of the "good" region. The maximum phase voltage was chosen to match the capability of a standard 3-phase bridge rectifier module, with the dc terminals short-circuited to form a 3-pulse forward-reverse cycloconverter, as shown in Figure 11.4. By using a 2-phase design, the number of modules and the control circuit complexity was reduced. Also, the motor was sized to operate directly from the converter without a filter.

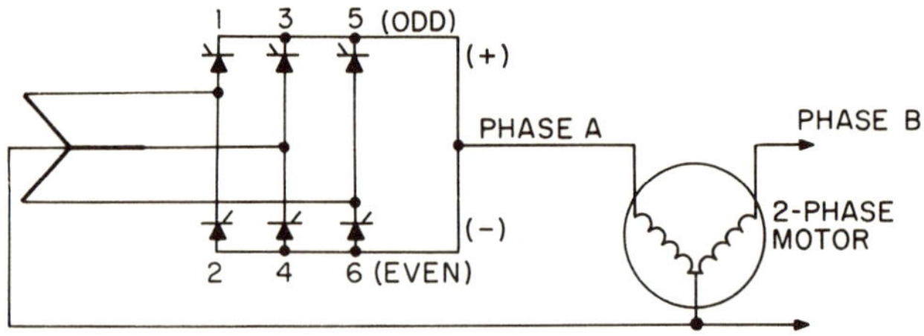

Figure 11.4. Cycloconverter for 13 Hz motor drive. Phase B is similar to phase A.

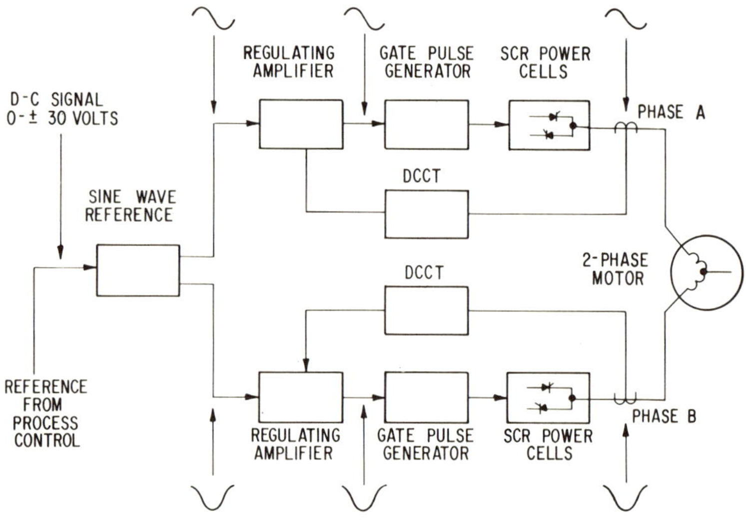

Figure 11.5. Block diagram of 2-phase motor drive system.

A block diagram of the drive control system is shown in Figure 11.5. The sine wave reference produces a pair of signals 90° apart in response to a bipolar dc reference signal from the process controller. The generated frequency is proportional to the magnitude of the dc signal, while the phase rotation is determined by the dc polarity. Amplitude of the sine waves is a function of the rate of change of the input signal.

These 2-phase sine wave reference signals are proportional to the instantaneous current desired in each phase of the motor load to meet the torque requirements of the drive. The actual phase currents are monitored by means of direct current current transformers (DCCTs), which are needed because the drive must reverse through zero frequency. In the current regulating amplifier, the reference and feedback signals are compared to form an error signal, which is the control signal input to the gate pulse generator.

The gate pulses for the thyristors in the SCR power circuit are produced at the intersections of the control signal with 60 Hz cosine firing waves, such as illustrated in Figures 5.1 and 5.2. Current polarity control logic determines which group of thyristors receives firing signals. In this particular system, there is a relatively long deadband interval with no current in the neighborhood of zero crossings, which is tolerable for the low output frequency.

Protection against faults or excessive overloads is obtained by suppressing the gate firing pulses when such a condition is sensed. This function is provided by the regulating amplifiers. In case of a commutation failure in the inverting mode, extinction of current in the thyristor that was conducting at the time of suppression may not occur until the motor emf reverses polarity.

In this particular installation, there are 14 table sections, each having a separate adjustable frequency power supply that is sized to match the duty cycle of the motors in that section. The continuous rating of the power supplies ranges from 20 kVA to 110 kVA, with 2 to 6 power thyristor modules in parallel to provide the needed capacity. The supply transformers have two primary and two secondary windings, one pair for each output phase. Although wound on the same core, they were designed to have low coupling to reduce interference between the two phases. The secondary windings have a zig-zag connection to eliminate the magnetic saturation effect of dc components in the windings.

An installed lineup of adjustable frequency cycloconverter power supplies is shown in Figure 11.6. This type of modular packaged equipment was typical of industrial power semiconductor rectifiers at the time of that installation. One of the trays is shown drawn out for inspection. The thyristors are cooled by forced air entering from the bottom and discharged at the top.

11.3 INDIVIDUAL AC MOTOR DRIVE SYSTEM

A very large cycloconverter has recently (1969) been installed to supply the motor driving a gearless grinding mill. The basic features of the system are outlined; more details are available in References

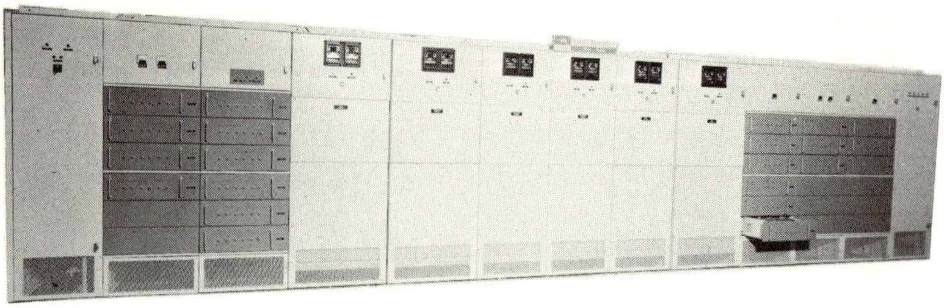

Figure 11.6. Lineup of cycloconverter power supplies for industrial application.

68 to 70 and 73. Grinding mills are used for pulverizing various materials, such as iron ore, the raw material for a cement kiln, and the clinker produced by the kiln. The latter case, termed a cement mill, is the application of this first gearless drive.

Essentially, a horizontal tube is supported by journal bearings at each end and is rotated slowly (12 to 20 r/min). The tube is clad with steel plates and may contain steel balls; the raw material is fed in at one end and is ground to a powder as it gravitates toward the other end, where it exits. Previously, the tube has been driven by gearing of some type, but mechanical problems arise as the size is increased. For a drum 5-m in diameter and 16.5-m long, requiring 8700 hp (6400 kW), a gearless drive was designed, using a synchronous motor of annular construction having a bore of 8-m.

The rotor assembly is bolted to a flange on the drum, while the stator construction is similar to that of a hydroelectric generator. With 44 poles, a frequency of 5.5 Hz is required for the maximum speed of 15 r/min. Since the supply frequency is 50 Hz in this case, the output/input frequency ratio is 0.11, well within the "good" waveform region for a 6-pulse converter, Figure 5.12.

Each phase of the stator is supplied by a 3-phase bridge cycloconverter, similar to the circuit shown in Figure 1.2. At the time when the equipment was designed, suitable high-voltage thyristors were not available, so mercury-arc rectifiers were employed. There are three 100 A (average), 3.5 kV (peak) rectifiers in parallel in each leg of the circuit. In future installations of this type, thyristors will probably be used. Separate supply transformers are provided for each output phase, allowing the motor windings to be connected in wye so that third harmonic currents cannot flow, even though the converter voltages are trapezoidal at high modulation levels. The converters are operated in the mode without intergroup circulating currents.

As in the drive system described in the previous section, the cycloconverter has a current-regulating inner loop. The 3-phase sinusoidal current reference signals are constructed to have the desired amplitude, frequency, and phase, such that control of the angle between the stator mmf vector and the direct axis of the rotor is achieved. For this, the speed and position of the rotor are measured by means of a special transducer generator. The rotor field excitation current is controlled as a function of the stator current magnitude by means of a thyristor-controlled rectifier. In this manner, the motor is adjusted to operate with its rated magnetic flux and present a unity power factor load to the cycloconverter, for all loads and speeds.

With this type of control, the torque is proportional to the stator current at all speeds, including standstill, and is not limited to a pull-out value. The drive system has characteristics similar to a dc motor drive. A speed-regulating outer loop is employed, in which the speed error is proportional to the torque or stator current desired. Thus, the dc signal representing the speed error is the input to the current reference generator.

The system is designed to accelerate the motor and its load with up to 160% of rated torque from standstill to a normal operating speed in the range 13 to 15 r/min. The starting current is also limited to about 160% of its rated value, much less than would be required without a controlled frequency converter. The voltage developed by the cycloconverter adjusts itself to produce the current demanded by the regulator. Under starting conditions, the modulation factor is low. However, at operating speed, the voltage is forced to become flat-topped to produce the higher fundamental component required. This, together with operation at unity motor power factor, allows the system power factor to be 0.86 or higher in the rated speed range.

Conventional mills often require auxiliary drives to turn the mill slowly to some desired position ("inching") for maintenance purposes. Here, this can be done by supplying a very low frequency from the cycloconverter. By applying dc to the motor, the mill can be held locked in position.

This system has features typical of a high-performance adjustable frequency drive for a synchronous motor. Similar high-performance drives for induction motors can be achieved by controlling the slip frequency in place of the torque angle of a synchronous motor. A signal proportional to the desired slip is added to the measured rotor speed signal to yield a measure of the frequency that must be generated by the converter.

11.4 COMPETITIVE SYSTEMS

Systems employing cycloconverters must pass scrutiny for technical and economic justification. As an electrical power frequency changer, a cycloconverter must compete with (1) rotating motor-generator sets and (2) dc link converters.

(1) Rotating motor-generator sets. Static power converters have the advantages of less expensive installation, much less maintenance, and less noise and vibration. However, the initial cost may be higher.

(2) dc Link converters (rectifier and inverter in cascade). Cycloconverters perform the conversion in only one stage and therefore should be more efficient and smaller in size, particularly because the dc filter is eliminated and the output waveform can be shaped rather easily. Further savings are realized if the inverter must be force-commutated. Thus, a cycloconverter is generally preferred, unless one of the following factors must be considered: A cycloconverter has a definite frequency ratio limitation, which an inverter does not have. In smaller sizes, fewer controlled switches (thyristors) may be required for an inverter, and an inexpensive diode rectifier might be used. If battery support during power outages is a system requirement, a dc link converter is the obvious choice. The input power factor with a rectifier link is generally higher, but the advantage depends upon whether the dc voltage is reduced by phase delay and the degree of modulation required of the cycloconverter.

When the application does not necessarily require frequency conversion, other alternatives must be weighed. In an electric drive system, there is the possibility of an ac system with rotating converters, such as the Krämer drive, as well as the dc Ward-Leonard motor-generator system. If these are rejected in favor of a static converter, then a choice between a dc and an ac drive must be made. A dc converter requires fewer controlled switches (thyristors), and it has a simpler control than a cycloconverter. At present, the cost of a converter is considerably greater than the cost of a motor. Hence, a dc drive is usually more economical, despite the higher cost and maintenance associated with the commutator.

However, as described in Sections 11.2 and 11.3, there are special cases where a cycloconverter ac drive has shown favor. These cases involve multiple motor drives or very large drives (the problems of a dc motor increase with size), and further application in these areas is anticipated with a widening range as the relative cost of thyristors improves. When an ac drive has been selected, the choice between a cycloconverter and a dc link converter, considered previously, must again be made.

In a drive system that does not have to be electric, cycloconverters or other electric converters must show advantages over mechanical means of speed control. The areas of aircraft generating systems (Section 11.1) and vehicular propulsion systems are examples where strong competition from nonelectric drives exists. At present, electric drives are in a small minority, but future developments are expected to change the balance.

In power conversion systems where the source is dc and the desired output is ac, a direct inverter is, usually, the obvious choice. However, if the output is at a low frequency and a transformer is necessary for voltage matching or isolation, the size of the transformer may be objectionable, or even prohibitive in some cases. For such applications, the cascade of a high-frequency inverter, small high-frequency transformer, and a cycloconverter has been proposed.[45,47] The arrangement is sometimes termed a "cycloinverter."[67] Forced commutation of the inverter provides natural commutation for the cycloconverter.

The inefficiency, complexity, and expense of an inverter-cycloconverter cascade system are severe handicaps. Justification is achieved only if the size of the transformer is prohibitive rather than merely objectionable. This assumes that the extra converter is smaller than the savings in transformer size. One application where the system might be suitable is for driving high-power ac electric locomotives supplied from a high voltage dc catenary.[79] Distribution voltages of 10 kV dc or more have been proposed, whereas it is difficult to design a traction motor winding for more than a few kilovolts. Since very low frequencies are required for starting, a transformer would be very large. The high-frequency link system offers a possible solution.

Reduction of transformer size has also prompted the suggestion of ac to ac power conditioning systems using the cascaded chain: diode rectifier, high-frequency inverter, small high-frequency transformer, and cycloconverter. With three stages of power conversion, this scheme suffers from greater inefficiency, complexity, and expense than the arrangement in the previous paragraph. It is doubtful whether converters of this type can gain any overall practical advantage.

Definitions of Rectifier Circuit Terminology

The following definitions are quoted from "Standard for Semiconductor Rectifier Components" AIEE No. 59, December 1962.

1.304 Single-way Rectifier Circuit

A single-way rectifier circuit is a circuit in which the current between each terminal of the alternating-voltage circuit and the rectifier circuit element conductively connected to it flows only in one direction. (See Note under 1.305.)

1.305 Double-Way Rectifier Circuit

A double-way rectifier circuit is a circuit in which the current between each terminal of the alternating-voltage circuit and the rectifier circuit elements conductively connected to it flows in both directions.

Note: The terms single-way and double-way provide a means for describing the effect of the rectifier circuit on current flow in transformer windings connected to rectifier circuits. Most rectifier circuits may be classified into these two general types. Double-way circuits are also referred to as bridge circuits.

1.310 Commutation

Commutation is the transfer of unidirectional current between rectifier circuit elements which conduct in succession.

1.311 Commutating Group

A commutating group is a group of rectifier circuit elements and the alternating-voltage supply elements conductively connected to them in which the direct current of the group is commutated between individual elements which conduct in succession.

1.312 Set of Commutating Groups

A set of commutating groups consists of two or more commutating groups which have simultaneous commutations.

1.313 Commutating Voltage

The commutating voltage is the phase-to-phase ac voltage of a commutating group.

1.314 Commutating Angle*

The commutating angle is the angle during which the current is commutated between two rectifier circuit elements.

1.315 Phase Control

Phase control is the process of varying the point within the cycle at which forward conduction is permitted to begin.

Note: The amount of phase control may be expressed in two ways:
(1) the reduction in dc voltage obtained by phase control, or
(2) the angle of retard, or advance.†

1.316 Commutating Reactance

Commutating reactance is the reactance which effectively opposes the transfer of current between rectifier circuit elements of a commutating group, or set of commutating groups.

Note: For convenience, the reactance from phase to neutral, or one half of the total reactance in the commutating circuit, is the value usually employed on computations, and is the value designated as the commutating reactance.

1.319 Simple Rectifier Circuit

A simple rectifier circuit is a rectifier circuit consisting of one commutating group if single-way or two commutating groups if double-way.

1.320 Parallel Rectifier Circuit

A parallel rectifier circuit is a rectifier circuit in which two or more simple rectifier circuits are connected in such a way that their direct currents add and their commutations coincide.

1.321 Series Rectifier Circuit

A series rectifier circuit is a rectifier circuit in which two or more simple rectifier circuits are connected in such a way that their dc voltages add and their commutations coincide.

1.322 Multiple Rectifier Circuit

A multiple rectifier circuit is a rectifier circuit in which two or more simple rectifier circuits are connected in such a way that their direct currents add, but their commutations do not coincide.

1.323 Cascade Rectifier Circuit

A cascade rectifier circuit is a rectifier circuit in which two or more simple rectifier circuits are connected in such a way that their dc

*Also termed the angle of overlap.
†Also termed delay angle or firing angle.

voltages add but their commutations do not coincide.

1.324 Number of Rectifier Circuit Phases

The number of phases in a rectifier circuit is equal to the total number of successive, nonsimultaneous commutations occurring within that rectifier circuit during each cycle when operating without phase control. It is also equal to the order of the principal harmonic in the direct voltage.

Note: The number of rectifier circuit phases influences both alternating-current and direct-current circuit wave form. In a simple single-way rectifier circuit, the number of rectifier circuit phases is equal to the number of rectifier circuit elements.

1.325 Mode of Operation

The mode of operation of a rectifier circuit is the characteristic pattern of operation determined by the sequence and duration of commutation and conduction.

Note: Most rectifier circuits have several modes of operation which may be identified by the shape of the current waves. The particular mode obtained at a given load depends upon the circuit constants.

The definitions below are quoted from IEC Publication 84 (1957).

130–1 Commutating Group

A group of discharge paths and phases which commutate independently of the others. Such a group shall include all windings whose commutations are simultaneous.

130–2 Commutating Number

The number of commutations occurring during one cycle in each commutating group.

130–3 Pulse Number

A characteristic of a converter connection expressed as the number of nonsimultaneous commutations occurring during one cycle in the converter.

Note: The pulse number is the same as the "number of rectifier circuit phases" defined in AIEE No. 59.

Glossary

A	load current, rms value, amperes.
A_D	load current in delta-rectifier circuit (delta load), rms value, amperes.
A_m	load current, crest value, amperes.
A_{m1}	load current for which regulation factor is specified, crest value, amperes.
A'_m	supply line current, maximum crest value of fundamental component, amperes.
A_Δ	circulating current in delta-rectifier circuit, maximum peak-to-peak amperes.
A_1	pseudo-instantaneous crest value of sine component of principal harmonic voltage as a fraction of E_{d0}.
A'_1	simultaneous value of A_1 for another output phase.
A_{1i}	value of A_1 at the ith selected time instant.
A_{1m}	current in delta-rectifier circuit, peak fundamental component, amperes.
$A_{11, 12, 13}$	value of A_1 for delta-rectifier circuit phases 1, 2, 3, respectively.
a	load current, instantaneous value, amperes.
a_m	crest value of sine component of mqth harmonic voltage.
a_1	crest value of sine component of qth harmonic voltage.
$a_{1,2,3}$	current in delta-rectifier circuit, instantaneous value in phases 1, 2, 3, respectively.
B_1	pseudo-instantaneous crest value of cosine component of principal harmonic voltage as a fraction of E_{d0}.

B_1'	simultaneous value of B_1 for another output phase.
B_{1i}	value of B_1 at the ith selected time instant.
$B_{11,12,13}$	value of B_1 for delta-rectifier circuit phases 1, 2, 3, respectively.
b_m	crest value of cosine component of mqth harmonic voltage.
b_1	crest value of cosine component of qth harmonic voltage.
C_d	snubber capacitance across thyristor, farads.
C_d'	snubber capacitance phase-to-phase, farads.
C_{1e}	rms value of pseudo-instantaneous crest value of principal harmonic voltage as a fraction of E_{d0}.
C_{1e}'	value of C_{1e} in delta-rectifier circuit with displaced-phase supplies.
c	duration of ramp portion of trapezoidal waveform, as a fraction of total period.
c_m	crest value of mqth harmonic voltage.
c_1	crest value of qth harmonic voltage.
c_{1e}	rms value of pseudo-instantaneous crest value of principal harmonic voltage.
(di/dt)	rate of change of current, amperes per second.
(dv/dt)	rate of change of voltage, volts per second.
E_a	total dc voltage drop due to thyristors, volts.
E_b	back emf in equivalent circuit of rectifier, volts.
E_c	commutating voltage at time of commutation, volts.
E_{cm}	commutating voltage, phase-to-phase, crest value, volts.
E_d	direct voltage output of rectifier, average value, volts.
E_{d0}	value of E_d with no firing delay and negligible overlap and losses.

$E_{d\alpha}$ value of E_d with firing delay α and negligible overlap and losses.

E_e supply phase voltage, line-to-neutral, rms value, volts.

E_m supply phase voltage, line-to-neutral, crest value, volts.

E'_m output voltage, peak instantaneous value, volts.

E_r dc voltage drop due to source resistance, volts.

E_x dc voltage drop due to commutation overlap, volts.

e supply phase voltage, line-to-neutral, instantaneous value, volts.

e_c commutating voltage, phase-to-next-phase, instantaneous value, volts.

e_i value of e for ith supply phase.

$e_m(\omega t)$ instantaneous value of mqth harmonic voltage, volts.

e_1 value of e for supply phase No. 1.

F output frequency, hertz.

f supply frequency, hertz.

I_d direct load current, average value, amperes.

I_e supply line current, rms value, amperes.

I'_e supply line current on primary side of transformer, rms value, amperes.

I_s current that saturates saturating reactors, amperes.

I_{1m0} supply line current, crest value of fundamental component, amperes.

i ordinal number.

i_1 supply line current, instantaneous value of fundamental component, amperes.

J_e thyristor current, rms value, amperes.

J_{e0}	thyristor current, rms value neglecting overlap, amperes.
J_m	thyristor current, peak value, amperes.
J_n	nth order Bessel function of the first kind.
J_{s1}	thyristor current, peak 1-cycle surge value.
J_{sr}	thyristor current, peak repetitive surge value.
J_0	thyristor current, average value, amperes.
J_1	first order Bessel function of the first kind.
j	ordinal number of thyristor firing.
K_a	numerical factor in Equation 9.7.
K_{c1}	surge crest factor relative to steady-state component.
K_{c2}	surge crest factor relative to steady-state component with negligible resistance.
K_f	voltage reduction factor caused by frequency-ratio effect.
K_n	parameter defined in Equation 2.38.
K_s	numerical factor in Equation 9.7.
$K\Delta$	voltage distortion factor caused by frequency-ratio effect.
k	modulation fraction of control signal.
L_c	commutating inductance, henry.
L_e	effective inductance in series with snubber, henry.
L_f	output filter inductance, henry.
L_g	intergroup inductance, henry.
L_p	source inductance, henry per phase.
L_s	saturated inductance of saturating reactor, henry.
L_u	unsaturated inductance of saturating reactor, henry.
L_0	effective inductance in equivalent circuit, henry.

L_0'	effective inductance in equivalent circuit of delta-rectifier configuration, henry.
M_g	mutual inductance of intergroup reactor, henry.
m	ordinal number of output cycle.
N	quality number of output waveform.
N_1	number of pairs of full steps in thyristor voltage waveform, per cycle.
N_2	number of pairs of half steps in thyristor voltage waveform, per cycle.
n_s	number of thyristors connected directly in series string.
n_w	number of circuit ways (1 for single way, 2 for double way).
n_0	number of output phases.
P_r	pseudo-instantaneous real power input to cycloconverter.
P_R	average real power input to cycloconverter.
P_t	total power factor of windings supplying a simple rectifier.
P_{t0}	value of P_t neglecting losses.
P_x	pseudo-instantaneous reactive power input to cycloconverter.
P_X	average reactive power input to cycloconverter.
P_1	average displacement power factor of cycloconverter.
p	commutating number of circuit.
p_r	real power coefficient.
p_x	reactive power coefficient.
q	pulse number of circuit.
R_c	equivalent source resistance, ohms per phase.
R_d	snubber damping resistance, ohms.
R_s	voltage equalizing resistance for series string of thyristors, ohms.

R_0	effective resistance in equivalent circuit, ohms.
R'_0	effective resistance in equivalent circuit of delta-rectifier configuration, ohms.
r	product of frequency ratio and modulation fraction.
s	number of commutating groups in series or cascade.
t	time, seconds.
t_i	time instant selected by numerical integration routine.
t_0	thyristor turn-off time.
U_a	thyristor conduction drop, volts.
U_e	thyristor voltage, rms value, volts.
U_m	thyristor voltage, peak value, volts.
U_s	magnitude of step change in thyristor voltage waveform, volts.
u	angle of overlap, radians.
V	load terminal voltage, rms value, volts.
V'_b	back emf in equivalent circuit of delta-rectifier cycloconverter, rms value of fundamental component, volts.
V_b	back emf in equivalent circuit of cycloconverter, rms value of fundamental component, volts.
V'_{bm}	crest value of V'_b.
V_D	voltage phasor defined by Equation 4.28.
V_J	voltage phasor defined by Equation 4.27.
V_{m0}	pseudo-crest value of output voltage neglecting overlap and losses.
V_0	pseudo-rms value of output voltage neglecting overlap and losses.
V'_0	value of V_0 including reduction factor for high frequency ratio.

V_0''	value of V_0 including reduction and distortion factors for high frequency ratio.
$V_{\Delta a}$	component of distortion voltage that leads V_0 by 90°, rms value.
$V_{\Delta b}$	component of distortion voltage in phase with V_0, rms value.
$V_{\Delta r}$	component of distortion voltage caused by variation of equivalent resistance, lags V_0 by 90°, rms value.
$V'_{\Delta r}$	value of $V_{\Delta r}$ in delta-rectifier circuit.
v_c	control signal voltage, instantaneous value, volts.
$v_{CB,AC,BA}$	pseudo-instantaneous terminal voltages of delta-rectifier circuit.
v_{f1}	sinusoidal firing wave voltage, instantaneous value.
v_{f2}	linear firing wave voltage, instantaneous value.
v_x	pseudo-instantaneous voltage drop due to commutation.
v_0	pseudo-instantaneous output voltage neglecting overlap and losses.
v_0'	value of v_0 including reduction factor for high frequency ratio.
$\bar{v}_0$	value of v_0 as a fraction of E_{d0}.
v_{0n}	desired pseudo-instantaneous output voltage from negative current group.
v'_{0n}	actual pseudo-instantaneous output voltage from negative current group.
v_{0p}	desired pseudo-instantaneous output voltage from positive current group.
v'_{0p}	actual pseudo-instantaneous output voltage from positive current group.
v_{0p1}	value of v_{0p} with modulation fraction of unity.
v_Δ	pseudo-instantaneous distortion voltage caused by high frequency ratio.

$v_{\Delta r}$	distortion component of v_x caused by high frequency ratio.
$v_{1,2,3}$	pseudo-instantaneous ideal output voltage of delta-rectifier phases 1, 2, 3, respectively.
W_s	average value of total thyristor snubber loss, watts.
W_{sm}	maximum value of total thyristor snubber loss, watts.
W_s'	average value of total phase-to-phase snubber loss, watts.
W_r	power loss in a voltage equalizing resistor, watts.
W_0	output power with unity power factor load, watts.
w_s	power loss in a thyristor snubber resistor, watts.
X_0	effective reactance in equivalent circuit, ohms.
x	parameter of definite integral.
x_i	ith value of x selected by numerical integration routine.
α	thyristor firing angle, radians.
$\alpha_{i1,2,3}$	firing angles at time t_i for delta-rectifier circuit phases 1, 2, 3, respectively.
α_n	thyristor firing angle, negative current group.
α_{ni}	thyristor firing angle, negative current group, at time t_i.
α_{ni}'	simultaneous value of α_{ni} for another output phase.
α_p	thyristor firing angle, positive current group.
α_{pi}'	simultaneous value of α_{pi} for another output phase.
α_r	thyristor firing angle retard limit.
α_0	thyristor firing angle advance limit.
α_1	thyristor firing angle, commutating group 1.
α_2	thyristor firing angle, commutating group 2.
β	$\sin^{-1}(E_d/E_m')$
γ	commutation margin angle, radians.

δ	phase displacement angle between sets of supply phases, radians.
ϵ	voltage regulation factor.
ζ	damping factor of snubber circuit
θ	angle between successive firings, radians at frequency ω.
θ_{gm}	geometric mean value of θ.
λ	phase displacement between output voltage and phase 1 of the supply, at time zero.
ξ	phase displacement angle of control signal from selected time origin.
π	3.14159265
σ	per unit snubber loss.
τ	thyristor voltage fall time constant at turn-on.
Φ	lag angle of load current with respect to control voltage.
Φ'	lag angle of load current with respect to terminal voltage.
Φ_{Δ}	difference between Φ and Φ'.
ϕ	lag angle of supply current with respect to phase voltage.
$\psi(\alpha,u)$	rms current function.
ψ	control angle of cycloconverter, defined by Equation 4.6.
Ω	angular output frequency, radians per second.
ω	angular supply frequency, radians per second.

Selected References and Bibliography

BOOKS (by date)

1. Prince, D.C., and Vogdes, F.B. *Mercury Arc Rectifiers and Circuits*, New York: McGraw-Hill Publishing Company, 1927.

2. Marti, O.K., and Winograd, H. *Mercury Arc Power Rectifiers*, New York: McGraw-Hill Publishing Company, 1930.

3. Glaser, A., and Müller-Lübeck, K. *Theorie der Stromrichter, Band 1*, Berlin: Julius Springer, 1935.

4. Rissik, H. *Mercury Arc Current Converters*, London: Sir Isaac Pitman & Sons, 2nd ed., 1941.

5. Adamson, C., and Hingorani, N.G. *High Voltage Direct Current Power Transmission*, London: Garraway, 1960.

6. Bedford, B. D., and Hoft, R.G. *Principles of Inverter Circuits*, New York: John Wiley & Sons, 1964.

7. Schaefer, J. *Rectifier Circuits: Theory and Design*, New York: John Wiley & Sons, 1965.

8. Möltgen, G. *Thyristoren in der Technischen Anwendung, Band 2: Netzgeführte Stromrichter*, Berlin: Siemens & Company, 1967.

9. Pelly, B. R. *Thyristor Phase-Controlled Converters and Cycloconverters*, New York: John Wiley & Sons, 1971.

RECENT CYCLOCONVERTER LITERATURE (since 1959)

This list is in approximate order of first publication. Only articles concerning cycloconverters, at least in part, are included. Many other papers dealing with ac/dc converters, particularly of the reversing type, contain information of interest in cycloconverter design.

10. Chirgwin, K. M., and Stratton, L. J. "*Variable-Speed, Constant-Frequency Generator System for Aircraft*," *AIEE Trans. Appl. Ind.* vol. 78, pt. II, pp. 304–310, 1959.

11. Jessee, R. D., and Spaven, W. J. "*Constant-Frequency AC Power Using Variable Speed Generation*," *AIEE Trans. Appl. Ind.*,

vol. 78, pt. II, pp. 411–418, 1959.

12. Chirgwin, K. M., Stratton, L. J., and Toth, J. R. "*Precise Frequency Power Generation from an Unregulated Shaft,*" *AIEE Trans. Appl. Ind.*, vol. 79, pt. II, pp. 442–451, 1960.

13. Caldwell, S. C., Peaslee, L.R., and Plette, D. L. "*The Frequency Converter Approach to a Variable Speed Constant Frequency System,*" *AIEE Conf. Paper*, CP 60–1076, August 1960.

14. Anonymous, "*AC Electric Drive,*" *Gas Turbine Publications*, p. 24, March–April 1961.

15. Griffith, D. C., and Ulmer, R. M. "*A Semiconductor Variable Speed AC Motor Drive,*" *Elec. Eng.*, pp. 350–353, May 1961.

16. Clarke, P. W. "*High Frequency Power Conversion (PNPN High to Low Frequency Converter),*" *AIEE Conf. Paper*, CP 62–335, January 1962.

17. Jessee, R. D., and Ernsberger, G. W."*Progress in the Development of a Variable-Speed, Constant-Frequency Electric Power System,*" *IEEE District Conf. Paper*, DP 63–781, March 1963.

18. VanEck, R. A. "*Frequency-Changer Systems Using the Cycloconverter Principle,*" *IEEE Trans. Appl. Ind.*, pp. 163–168, May 1963.

19. Hoolboom, G. J. "*A Polyphase, All Solid-State Cycloconverter,*" *IEEE Conf. Paper*, CP 63–1040, October 1, 1963.

20. Heck, R., and Meyer, M. "*A Static Frequency-Changer-Fed Squirrel-Cage Motor Drive for Variable Speed and Reversing,*" *Siemens Rev.* No. 11, pp. 401–405, November 1963.

21. Plette, D. L., and Carlson, H. G. "*Performance of a Variable-Speed Constant Frequency Electrical System,*" *IEEE Trans. Aerospace*, vol. AS–2, pp. 957–970, April 1964.

22. Faust, W. "*Static Frequency Changers for 16⅔ c/s Railway Networks,*" *Brown Boveri Rev.*, vol. 51, No. 8/9, pp. 519–525, 1964.

23. Schönung, A. "*Varying the Speed of Three-Phase Motors by Means of Static Frequency Changers,*" *Brown Boveri Rev.*, vol. 51, no. 8/9, pp. 540–554, 1964.

24. Guyeska, J. C., and Jordan, H. E. "*Cycloconverter Adjustable Frequency Drives*," *IEEE Textile Ind. Conf.*, October 1–2, 1964.

25. Guyeska, J. C., and Jordan, H. E. "*Static AC Variable Frequency Drive*," *Proc. Nat. Elect. Conf.*, vol. 20, 1964, pp. 358–365.

26. Brown, M., and Harvery, W. T. "*A Variable Frequency Supply for Induction Motors*," *Proc. Conf. Appl. Large Ind. Drives*, London, England, pp. 120–123, May 5–7, 1965.

27. Borland, R. P., and Lawson, L. J. "*An AC Electric Drive System for Commercial Applications*," *Cleveland Electronics Conf.*, April 1965.

28. Borland, R. P., and Puchy, C. G. "*An AC Electric Antenna Drive System for Aerospace Applications*," *Proc. IEEE Nat. Aerospace Elect. Conf.*, pp. 144–149, May 10–12, 1965.

29. Amato, C. J., and Diczhazy, R. B. "*Lower Order Distortion Terms in Frequency Converters*," *Proc. IEEE Nat. Aerospace Elect. Conf.*, pp. 165–172, May 10–12, 1965.

30. Lawson, L. J., "*Static Frequency Converter*," *Proc. 19th Ann. Power Sources Conf.*, pp. 135–137, May 18–20, 1965.

31. Chirgwin, K. M. "*A Variable-Speed Constant Frequency Generating System for a Supersonic Transport*," *Suppl. IEEE Trans. Aerospace*, pp. 387–392, June 1965.

32. Lawson, L. J. "*Precisely Controlled Three-phase Squirrel Cage Induction Motor Drives for Aerospace Applications*," *Suppl. IEEE Trans. Aerospace*, pp. 93–97, June 1965.

33. Amato, C.J. "*Sub-Ripple Distortion Components in Practical Cycloconverters*," *Suppl. IEEE Trans. Aerospace*, pp. 98–106, June 1965.

34. Amato, C. J. "*Analog Computer Simulation of an SCR as Applied to a Cycloconverter*," *Proc. Nat. Elec. Conf.*, vol. 21, pp. 933–937, 1965; (or *IEEE Trans. Ind. Gen. Appl.*, vol. IGA–2, pp. 137–140, March/April 1966).

35. Slabiak, W., and Lawson, L. J. "*Precise Control of a Three-phase Squirrel Cage Induction Motor Using a Practical Cycloconverter*," *Proc. Nat. Elec. Conf.*, vol. 21, pp. 938–943, 1965; (or *IEEE Trans. Ind. Gen. Appl.*, vol. IGA–2, pp. 274–280, July/August 1966).

36. Hamilton, R. A., and Lezan, G. R. "*Thyristor Adjustable Frequency Power Supplies for Hot Strip Mill Run-out Tables,*" *IEEE Pub. 34C20*, Record IEEE Industrial Static Power Conversion Conf., pp. 69–77, Nov. 1–3, 1965; (or *IEEE Pub. 34C36*, Conf. Record IGA 1st Annual Meeting, pp. 641–652, Oct. 3–6, 1966); (or *IEEE Trans. Ind. Gen. Appl.*, vol. IGA–3, pp. 168–175, March/April 1967).

37. Amato, C. J. "*Variable Speed with Controlled Slip Induction Motor,*" *IEEE Pub. 34C20*, Record IEEE Ind. Static Power Conversion Conf., pp. 181–185, Nov. 1–3, 1965.

38. Slabiak, W., and Lawson, L. J. "*Optimizing Control Systems for Land Vehicles,*" *IEEE Pub. 34C20*, Record IEEE Ind. Static Power Conversion Conf., pp. 186–189, Nov. 1–3, 1965.

39. Abraham, L., Forster, J., and Schliephake, G. "*AC Motor Supply with Thyristor Converters*" IEEE Pub. 34C20, Record IEEE Ind. Static Power Conversion Conf. pp. 210–216, Nov. 1–3, 1965; (or *IEEE Trans. Ind. Gen. Appl.*, vol. IGA–2, no. 5, pp. 334–340, Sept./Oct. 1966).

40. Bowler, P. "*The Application of a Cycloconverter to the Control of Induction Motors,*" Conf. Power Applications of Controllable Semiconductor Devices, London, England, *IEEE Pub. No. 17*, pp. 137–145, Nov. 10–11, 1965.

41. Slabiak, W. "*An AC Electric Individual Wheel Drive System for Land Vehicles,*" *S.A.E. Paper No. 660134*, S.A.E. Automotive Engineering Congress, January 1966; or *S.A.E. Trans.*, vol. 75, pp. 664–671, 1967.

42. Lawson, L. J., Borland, R. P., and Puchy, C. G. "*Optimal Control System Performance of an AC Electric Vehicular Wheel Drive,*" *S.A.E. Paper No. 660135*, S.A.E. Automotive Engineering Congress, January 1966; (or *S.A.E. Trans.*, vol. 75, pp. 672–680, 1967).

43. Amato, C. J. "*An AC Equivalent Circuit for a Cycloconverter,*" *1966 IEEE International Conv. Record*, vol. 14, pt. 8, pp. 24–30; (or *IEEE Trans. Ind. Gen. Appl.*, vol. IGA–2, pp. 358–362, Sept./Oct., 1966).

44. Lawson, L. J., and Amato, C. J. "*An AC Electric Drive System for Locomotive and Multiple-Unit Traction Systems,*" *IEEE Paper CP 66–204*. Presented at the Ninth Joint ASME/IEEE Railroad Conf., San Francisco, Calif., May 1–6, 1966.

45. Lawson, L. J. "*Static Inverters for Operation from Fuel Cells,*" *Proc. 20th Ann. Power Sources Conf.*, pp. 204–207, May 24–26, 1966.

46. Pinter, G. E. "*The Cycloconverter Adjustable Speed Drive,*" *Machine Design*, p. 150, June 23, 1966.

47. Lawson, L. J. "*Fuel Cell Power Conversion by the AC Link Type Static Inverter,*" 1966 Aerospace Systems Conf. Record, *Suppl. IEEE Trans. Aerospace and Electronic Systems*, vol. AES–2, no. 4, pp. 160–163, July 1966.

48. Lawson, L. J. "*An Ultra High Density Power Conversion System Package,*" *7th Internat. Elec. Circuit Packaging Symp. Record*, paper 7/22, Aug. 22–23, 1966.

49. Lawson L. J. "*Design Considerations for a Modern Cycloconverter,*" *Symposium on Power Semiconductor Technology*, Univ. of Missouri, Columbia, Mo., September 1966.

50. Lawson, L. J., and Slabiak, W. "*A Thyristor Drive for High Performance Land Vehicles,*" *Proc. Nat. Elec. Conf.*, vol. 22, pp. 271–276, 1966.

51. Reimers, E., and Amato, C. J. "*Performance Characteristics of a Controlled Slip Induction Motor Drive,*" (abstract) *IEEE Pub. 34C36*, Conf. Record IGA 1st Annual Meeting, p. 499, Oct. 3–6, 1966.

52. Lawson, L. J. "*The Practical Cycloconverter,*" *IEEE Pub. 34C36*, Conf. Record IGA 1st Annual Meeting, pp. 123–128, Oct. 3–6, 1966; (or *IEEE Trans. Ind. Gen. Appl.* vol. IGA–4, pp. 141–144, March/April 1968).

53. Schwartz, S. A., and Femling, D. "*Single-phase Control for Cycloconverter,*" *Fairchild Application Bulletin*, APP–138, August 1967.

54. Splatt, R. J. K. "*An Experimental Cycloconverter,*" (British) *Royal Aircraft Establishment Technical Report 67270*, October 1967.

55. Pudlo, T. "*SCR Simulation on a Hybrid Computer,*" *IEEE Pub. 34C62*, Conf. Record IGA 2nd Annual Meeting, pp. 39–43, Oct. 2–5, 1967.

56. Lawson, L. J. "*A True No-break, Off Line Uninterrupted Power Supply,*" *IEEE Pub. 34C62*, Conf. Record IGA 2nd Annual Meeting, pp. 154–158, Oct. 2–5, 1967.

57. Bland, R.J. "*Factors Affecting the Operation of a Phase-Controlled Cycloconverter,*" *Proc. IEE* (British), vol. 114, no. 12, pp. 1908–16, Dec. 1967.

58. Gyugyi, L. "*A Study on the Application of the Force-Commutated Cycloconverter for Variable Speed Control of AC Machines,*" M.S. thesis, University of Pittsburgh, 1967.

59. Yohman, E. J. "*Characteristics of SCR Cycloconverter Power Stage,*" *WESCON Tech. Papers,* vol. 12, pt. 4, Session 26/3, 1968.

60. Yamaguchi, M., and Fujiwara, H. "*Characteristics of Graetz-Connection Type Cycloconverter,*" *Elec. Eng.* (Japan), vol. 88, no. 10, pp. 28–37, Oct. 1968.

61. Miyairi, S., and Takahashi, I. "*Application of Power Modulation Technique Employing Thyristors,*" *Elec. Eng.* (Japan), vol. 88, no. 11, pp. 36–45, Nov. 1968.

62. Miyairi, S., and Shioya, M. "*Gate Control Circuit for Thyristor Cycloconverter,*" *Elec. Eng.* (Japan), vol. 88, no. 12, pp. 31–41, December 1968.

63. Rich, E.A.E. "*Concepts of Gearless Ball-Mill Drives,*" *IEEE Trans. Ind. Gen. Appl.,* vol. IGA–5, no. 1, pp. 13–17, Jan./Feb. 1969.

64. Jenkins, J. E. "*A Modified Cycloconverter for Use with High Frequency Sources,*" *IEE Pub. No. 53,* Part 1–Contributions, pp. 313–319,Conf. on Power Thyristors and their Applications, London, England, May 6–8, 1969.

65. Masterman, R. H. "*Cycloconverter Amplifier Allowing Wide Variation in Load Impedance,*" *Elec. Letters,* vol. 5, no. 14, pp. 306–308, July 10, 1969.

66. Jovy, R. K. "*Direktumrichter für die Frequenzerhöhung mit festem Ubersetzungsverhältnis*" (Direct converter for raising the frequency with a fixed step-up ratio), *VDE-Fachtagung Elektronik,* pp. 123–134, 1969.

67. Nishimura, M., Murakami, Y., Sakuma, N., and Kuroyama, T. "*A Pulsewidth-Controlled Cycloinverter,*" *Elec. Eng.* (Japan), vol. 89, no. 10, pp. 46–56, 1969.

68. Bläuenstein, E. "*The First Gearless Drive for a Tube Mill,*" *Brown Boveri Rev.,* vol. 57, no. 3, pp. 96–105, March 1970.

69. Langer, J. "*Static Frequency Changer Supply System for Synchronous Motors Driving Tube Mills*," *Brown Boveri Rev.*, vol. 57, no. 3, pp. 112–119, March 1970.

70. Stemmler, H. "*Drive System and Electronic Control Equipment of the Gearless Tube Mill*," *Brown Boveri Rev.*, vol. 57, no. 3, pp. 120–128, March 1970.

71. Tsuchiya, T. "*Basic Characteristics of Cycloconverter-Type Commutatorless Motors*," *IEEE Trans. Ind. Gen. Appl.*, vol. IGA–6, no. 4, pp. 349–356, July/August 1970.

72. Dewan, S. B., and Kankam, M. D. "*A Method for Harmonic Analysis of Cycloconverters*," *IEEE Trans. Ind. Gen. Appl.*, vol. IGA–6, no. 5, pp. 455–462, Sept./Oct. 1970.

73. Würgler, H. U. "*The World's First Gearless Mill Drive*," *IEEE Trans. Ind. Gen. Appl.*, vol. IGA–6, no. 5, pp. 524–527, Sept./Oct. 1970.

74. Hamblin, T. M., and Barton, T. H. "*Cycloconverter Control Circuits*," *IEEE Pub. 70C1–IGA*, Conf. Record of IGA 5th Annual Meeting, pp. 559–571, Oct. 5–8, 1970.

75. Long, W. F., and Schnitz, N. L. "*Cycloconverter Control of the Doubly Fed Induction Motor*," *IEEE Trans. Ind. Gen. Appl.*, vol. IGA–7, no. 1, pp. 95–100, Jan./Feb. 1971.

EARLY CYCLOCONVERTER LITERATURE

There is extensive literature, mostly in German, describing work on 50/16⅔ Hz cycloconverters for railroad electrification during the 1930s. Reference 76 is a bibliography of this work, Reference 77 is a classic paper on cycloconverters, and Reference 78 is a comprehensive English-language survey of the German work.

76. "Auslese aus dem Schrifttum in Zeitschriften über Bahnumrichter und Wechselrichter von 1932 an" (Selected bibliography on railroad converters and inverters since 1932), *Elektrische Bahnen*, vol. 13, no. 8/9, p. 229, 1937.

77. von Issendorff, J. "*Der gesteuerte Umrichter*" (The controlled converter), *Wissenschaftliche Veröffentlichungen aus den Siemens-Werken*, vol. 14, pt. 3, pp. 1–31, 1935.

78. Feinberg, R. "*Frequency Changing with Mercury Arc Mutators*," *JIEE.*, vol. 85, pp. 531–543, 1939.

RUSSIAN LITERATURE

79. Dolaberidze, G. P. "*Conversion of DC to Controlled-Frequency AC for Electric Traction*," *Elektrotekhnika*, vol. 36, no. 4, pp. 58–62, April 1965. (English trans. in *Soviet Electrical Engineering*, vol. 36, no. 4, pp. 84–90, 1965.)

80. Bernshtein, I. Ya. "*A Frequency Changer without a DC Section*," *Elektrotekhnika*, vol. 36, no. 6, pp. 38–42, June 1965. (English trans. in *Soviet Electrical Engineering*, vol. 36, no. 6, pp. 56–61, 1965.)

81. Ettinger, E. L., and Bernshtein, I. Ya. "*Compensating Currents in Thyristor Converters*," *Elektrotekhnika*, vol. 38, no. 3, pp. 5–6, March 1967.

82. Firago, B. I. "*Control System of Frequency Converter without Intermediate DC Circuit Element*," Izvestiya Vysshikh Uchebnykh Zavedenii, *Energetika*, no. 11, pp. 31–36, Nov. 1966.

83. Firago, B. I. "*Power Factor of Frequency Converters Having No DC Circuits*," *Energetika*, no. 8, pp. 41–47, Aug. 1967.

84. Firago, B. I. "*Harmonic Analysis of Voltage Characteristics for Frequency Converters Having No DC Circuits*," *Energetika*, no. 7, pp. 31–36, July 1968.

85. Djemerov, G. G. "*The Influence of Frequency Converters with Direct Coupling on their Supply Circuit*," *Elektrichestvo*, no. 4, pp. 24–29, April 1968.

86. Bikhov, U. M., Wouls, M. L., and Furman, V. B. "*A Study of the Load on the Valves in Frequency Converters without a DC Link*," *Elektrichestvo*, no. 1, pp. 72–76, Jan. 1970.

87. Djemerov, G. G. "*A Method for Analyzing the Output Voltage of Frequency Converters with Direct Coupling*," *Elektrichestvo*, no. 8, pp. 71–76, Aug. 1970.

88. Abramovitch, B. N., Biriukov, U. A., Vadaturski, V. M., Lagutin, V. I., and Glebov, I. A. "*Synchronous Machine Excitation Systems using Converters Conducting in Both Directions*," *Elektrichestvo*, no. 11, pp. 1–5, Nov. 1970.

Index